M-libraries 3

Transforming libraries
with mobile technology

M-libraries 3

Transforming libraries
with mobile technology

edited by
Mohamed Ally and **Gill Needham**

facet publishing

Published by Facet Publishing,
7 Ridgmount Street, London WC1E 7AE
www.facetpublishing.co.uk

Facet Publishing is wholly owned by CILIP: the Chartered Institute of Library and
Information Professionals.

British Library Cataloguing in Publication Data
A catalogue record for this book is available from the British Library.

ISBN 978-1-85604-776-0

First published 2012

Mixed Sources
Product group from well-managed
forests and other controlled sources
www.fsc.org Cert no. SA-COC-1565
© 1996 Forest Stewardship Council

FSC

Text printed on FSC accredited material.

Typeset from author's files by Flagholme Publishing Services in 12/14 American Garamond
and Nimbus Sans.
Printed and made in Great Britain by MPG Books Group, UK.

Contents

Acknowledgements.. ix

Contributors.. xi

Foreword...xxv
Stephen Abram

Introduction... xxix
Gill Needham

1   Education for all with mobile technology: the role of
    libraries...1
    Mohamed Ally

PART 1 DEVELOPING MOBILE SERVICES 11
2   Preparing for the mobile world: experimenting with changing
    technologies and applications for library services....................13
    Colin Bates and Rebecca Carruthers

3   Enhancing open distance learning library services with mobile
    technologies 23
    Jenny Raubenheimer

4   Use of mobile phones in the delivery of consumer health
     information 33
     Vahideh Zarea Gavgani

5   Deploying an e-reader loan service at an online university 43
     Anna Zuñiga-Ruiz and Cristina López-Pérez

6   Mobile service providers and library services in a
     multi-campus library 51
     Ela Volatabu Qica

7   Using mobile technology to deliver information in audio format:
     learning by listening...57
     Margie Wallin, Kate Kelly and Annika McGinley

8   Sound selection: podcasts prove positive 65
     Daniel McDonald and Roger Hawcroft

PART 2 PEOPLE AND SKILLS..73

9   Staff preparedness to implement mobile technologies in
     libraries...75
     Sarah-Jane Saravani and Gaby Haddow

10 Apps and attitudes: towards an understanding of the
     m-librarian's professional make-up 85
     Kate Davis and Helen Partridge

11 There's a librarian in my pocket: mobile information literacy
     at UTS Library..93
     Sophie McDonald

12 Exploring student engagement with mobile technologies.......101
     Julie Cartwright, Sally Cummings, Bernadette Royal,
     Michelle Turner and JoAnne Witt

13 It's just not the same: mobile information literacy...................109
     Andrew Walsh and Peter Godwin

14 The students have iPods: an opportunity to use iPods as a
teaching tool in the library 119
Iris Ambrose

PART 3 FOCUS ON TECHNOLOGY...127

15 Mobile services of the National Library of China.....................129
Wei Dawei, Xie Qiang and Niu Xianyun

16 India's mobile technology infrastructure to support
m-services for education and libraries.......................................139
Seema Chandhok and Parveen Babbar

17 Use and user context of mobile computing: a rapid
ethnographic study 151
Jim Hahn

18 Meeting the needs of library users on the mobile web...........159
Hassan Sheikh and Keren Mills

19 Mobile dynamic display systems for library opening
hours..171
Keiso Katsura

20 Device-independent and user-tailored delivery of
mobile library service content 181
*Damien Meere, Ivan Ganchev, Máirtín Ó Droma,
Mícheál Ó hAodha and Stanimir Stojanov*

21 Designing effective mobile web presence 191
Sam Moffatt

Conclusion..197
Mohamed Ally

Index...201

Acknowledgements

Without the Third International M-Libraries Conference this book would not exist, and so we would like to thank everyone who contributed to its success: the universities responsible for the conference (University of Southern Queensland (USQ), Athabasca University, Thompson Rivers and the Open University (UK)); the International Organizing Committee, Programme Committee and Local Organizing Committee; and all our excellent speakers, session chairs, delegates, helpers and commercial sponsors. We would like to give particular thanks to our colleagues, Nicky Whitsed, Nancy Levesque, Sue Craig, Marisa Parker, Steve Shafer and Paul Coyne, for their brilliance in pulling the conference together, and a special mention to Marisa and her team at USQ for hosting the event with such efficiency and style.

Our thanks go to all the authors who have contributed to the book, not only for their inspired content but for their co-operation in meeting deadlines. Above all, our thanks go to Jeannette Stanley for her patience, hard work and support.

Mohamed Ally
Gill Needham

Contributors

Keynote speakers

Stephen Abram, MLS, is past President 2008 of the Special Libraries Association and the past President of the Canadian Library Association. He is the Vice President for Strategic Partnerships and Markets for Gale Cengage. He was Vice President Innovation for SirsiDynix and Chief Strategist for the SirsiDynix Institute. He was Publisher Electronic Information at Thomson after managing several libraries. Stephen was listed by *Library Journal* as one of the top 50 people influencing the future of libraries. He has received numerous honours and regularly speaks internationally. His columns appear in *Information Outlook* and *Multimedia and Internet@Schools*, *OneSource*, *Feliciter* and *Access*, and he also writes for *Library Journal*. He is the author of ALA Editions' best-selling *Out Front with Stephen Abram*. His blog, Stephen's Lighthouse, is popular in the library sector.

Dr Mohamed Ally is Professor in Distance Education and Director of the Centre for Distance Education at Athabasca University, Canada. His current areas of research include the use of mobile technology in learning and libraries and e-learning. Dr Ally is past President of the International Federation of Training and Development Organizations (IFTDO) and is one of the Founding Directors of the International Association of Mobile Learning (IamLearn). He was also on the Board of the Canadian Society for Training and Development.

Dr Ally chaired the Fifth World Conference on Mobile Learning and co-chaired the First International Conference on Mobile Libraries. He recently edited three books on the use of mobile technology in education and libraries. In addition, Dr Ally has published articles in peer-reviewed journals and chapters in books and encyclopaedias and has served on many journal boards and conference committees. He has presented keynote speeches, workshops, papers and seminars in many countries.

Gill Needham is currently Associate Director (Information Management and Innovation) at the Open University Library. Her varied career has encompassed roles as researcher, librarian, project manager and educator. She has more than 25 years' experience in senior roles in the library and information sector, with an emphasis on innovation and skills development. During her 13 years at the Open University she has led strategic and developmental work on information literacy and has authored on a number of Open University courses, including Knowledge, Information and Care, Making Sense of Information in the Connected Age, Beyond Google and, most recently, the Evolving Information Professional. She was awarded a National Teaching Fellowship by the Higher Education Academy in 2006 and holds a Master's degree in Health Information Management and first degree in English and American Literature. She has published and presented widely throughout her career.

Contributors

Iris Ambrose is the Campus Librarian at La Trobe University's Shepparton campus in the Goulburn Valley, Victoria, Australia. The library is shared with the Goulburn Ovens Institute of TAFE (Technical and Further Education). The collection caters for 300 students from Commerce, Arts and Education faculties, while nursing students also use a shared library with Goulburn Valley Health. She has been in this role for ten years and is now also the campus Academic Language and Learning lecturer. She has recently completed a Master's degree in Information Management and Systems degree with Monash University, taking off-campus units. She won the 2009 VALA Student Award for excellent results.

Parveen Babbar is Assistant Librarian in the Library and Documentation Division at Indira Gandhi National Open University, New Delhi, India. Prior to this he was associated with the University of Delhi, India. He has more than eight years of professional experience. He holds an MPhil (Library and Information Science) from Annamalai University, MLISc from the University of Delhi and Master in Computer Application and Master in Business Administration from Indira Gandhi National Open University. He had been Gold Medallist in MLISc at Delhi University. He has been Junior Research Fellow and has passed the JRF-NET exam of UGC, India. He has participated in many projects in library automation and digitization, such as CALPI Library Setup for a Swiss agency. He is a member of the Special Libraries Association (SLA), USA, and has been the webmaster on the board of the Asian Chapter, SLA. He has presented many papers at national and international conferences and has published about 20 papers in journals and conference proceedings. He has been a member of the organizing committee of a number of national and international conferences related to library and information science.

Colin Bates is the Manager, Library Services, for the Faculty of Health and Geelong at Waurn Ponds Campus Library, at Deakin University. He is responsible for library services, programmes and resources supporting the faculty, and for local library operations at the Geelong at Waurn Ponds Campus, and is also responsible for online and off-campus library services and support. He has many years of experience in the provision of online and electronic library services and resources and the use of technology to support the use of library services and resources by on- and off-campus students. Recent initiatives with the School of Medicine have ensured the effective online and mobile information resources options are available to medical students undertaking remote clinical placements.

Rebecca Carruthers is the Collections Specialist for Monographs at Deakin University, Australia. Her responsibilities include purchasing print and e-books for all four campuses of Deakin University, as well as creating access to purchased and gratis e-books. Being a member of Net-Gen, Rebecca is keen to assess the benefit of including technology in any environment. She is currently responsible for several projects on e-book readers and mobile technology, including Kindle and iPad loan programmes.

Julie Cartwright, Sally Cummings, Bernadette Royal and **Michelle Turner** combine to form the Library Liaison team at Charles Darwin

University, Northern Territory, Australia. As a team, they are led by **JoAnne Witt**, Liaison Information and Literacy Co-ordinator. The team members have varied backgrounds in academic, vocational education and training, school, public, special and government libraries as well as in industry. They work together so well because they share an enthusiasm for delivering high-quality services to their clients and for developing and implementing new ideas and technologies in the library setting. Trainee librarian Ben Heaslip also contributed to this project.

Dr Seema Chandhok is Deputy Librarian at Indira Gandhi National Open University (IGNOU), New Delhi, India. She has more than 23 years' professional experience, including 18 years as Assistant Librarian at IGNOU. She has a Master's degree in Public Administration, Distance Education, Library and Information Science and a PhD in Library and Information Science. He work has been published in professional journals and conference proceedings and she has also published indexes to articles published in journals.

Kate Davis is an Associate Lecturer in Library and Information Science at the Queensland University of Technology (QUT). Kate began her career in libraries in 2004 as a Reference Librarian at QUT, where she worked while completing her studies in Library and Information Science. In 2005, Kate participated in the graduate programme at the National Library of Australia. She then worked in various roles, including serials cataloguing and reference. Most recently, Kate was Online Futures Librarian at Gold Coast City Council Libraries, where she was responsible for online collections and services. Her teaching and research interests centre round the application of technology to library service provision, particularly in the public library context, the delivery of information literacy training online and corporate use of social media. Kate is currently a PhD student and is investigating the information practices of new mothers in social media spaces.

Dr Ivan Ganchev is a Senior Member of the Institute of Electrical and Electronic Engineers (IEEE). He received his engineering and doctoral degrees from the Saint-Petersburg State University of Telecommunications in 1989 and 1994, respectively. He is an Associate Professor at the

University of Plovdiv and an ITU-T (International Telecommunication Union, Telecommunication Standardization Sector) Invited Expert. Currently he is lecturing in the University of Limerick (Ireland), where he is also a Deputy Director of the Telecommunications Research Centre. Previously he served as a member of the Academic Network for Wireless Internet Research in Europe (ANWIRE) and two European Science Foundation Cooperation in Science and Technology research (COST) Actions: 'Modelling and Simulation Tools for Research in Emerging Multi-service Telecommunications' and 'Traffic and QoS Management in Wireless Multimedia Networks' (COST 285 and 290). Currently he is a member of the COST Action IC0906, 'Wireless Networking for Moving Objects' (WiNeMO). His research interests include: simulation and modelling of complex telecommunication systems, new communications paradigms for wireless next-generation networks (NGN), always best connected and best served (ABC&S), third-party authentication, authorization and accounting (3P-AAA) management, wireless billboard channels (WBC), internet tomography and m-learning ICT. Dr Ganchev has served on the technical programme committees of a number of prestigious international conferences and workshops.

Dr Vahideh Zarea Gavgani is a researcher and medical information specialist and works as an assistant professor at Tabriz University of Medical Sciences, Department of Medical Library and Information Sciences, Tabriz, Iran. She has published 33 papers in international conferences and journals, three books in Persian and three book chapters (international contribution), and has translated one book into Persian. She is reviewer for IGI Global Publishing in the field of medical information science, a member of the international review board of the *International Journal of User Driven Health Care* (IJUDHC), an international committee member and reviewer for the 6th International Conference on Evidence Based Library and Information Practice (EBLIP6) and reviewer of the *Journal of Birjand University of Medical Sciences*. She is a member of the Iranian Center of Evidence Based Medicine, a teacher of the Training the Trainers programme in the RDCC (Research Development Centre), Tabriz University of Medical Sciences, and in the Online faculty of the Online Problem Solving Course (OPSC) in the Cal2Cal Institute, Indian Branch.

Peter Godwin is currently working at the University of Bedfordshire in Luton, UK. Formerly he was Academic Services Manager at London South Bank University, in charge of subject support to all faculties. His interest in information literacy has focused on support to academic staff in universities and the impact of Web 2.0 on literacy in all information sectors. He has presented widely on Web 2.0 and how this affects the content and delivery of information literacy. In 2008 he co-edited the pioneering book *Information Literacy meets Library 2.0* for Facet Publishing, which is supplemented by a blog of the same name. Currently he is investigating how mobile devices will affect library services. He draws on many years' experience in academic library management and has presented at conferences in Europe, Asia, the USA and Canada.

Gaby Haddow is a lecturer with the Department of Information Studies at Curtin University. She was appointed to the editorial board of the *Australian Library Journal* in 2010 and is a member of the Australian Library and Information Association's Research Committee. For three years Gaby wrote evidence summaries for the journal *Evidence Based Library and Information Practice* and she has published in national and international journals. She has conducted a number of small studies focused on scholarly communication and research assessment and on the communication of research to practice. Gaby's previous position as Humanities Faculty Librarian for Curtin Library has led to an ongoing collaborative project investigating library use and retention of commencing students.

Jim Hahn holds the position of Orientation Services Librarian and Assistant Professor of Library Administration at the University of Illinois at Urbana-Champaign. His day-to-day focus is on helping first-year and transfer students to make the transition to university study. His current research focus concerns understanding how library information resources can be provided to a student's hand-held device, based on position in the library stacks. Recently he has led prototyping research studies on the development of a mobile way-finding app.

Roger Hawcroft, DipT, BEd, MAppSc, AALIA, has over 30 years' experience in a variety of libraries, as well as teaching and commercial experience. His strength is a strong client focus and a willingness to challenge conventional approaches, to innovate and enthuse, and to encourage staff to implement creative solutions.

Keiso Katsura is a male professor of Library and Information Science at Miyagi Gakuin Women's University, Sendai City, Northern Japan. He was born in 1950 in Hiroshima Prefecture and educated at Keio University, Tokyo (BA Library and Information Science, 1973; MA Sociology, 1977) and Columbia University (MS Library Service, 1988). He worked as a librarian at the Japan International Cooperation Agency, Tokyo, from 1979 to 1992. He also worked as a researcher at the Hypermedia Laboratory, University of Tampere, Finland, during 2004/5. He is currently involved in university courses, delivering web and mobile-based lectures with a range of Web 2.0 content. His main area of interest is in the information search and retrieval literacy of academics.

Kate Kelly is Library Manager of Southern Cross University's Gold Coast and Tweed Heads Campus library. During her career, she has seen the library transform from fortress to hand-held. Her current research interest is in the opportunities that mobile devices and associated applications provide in teaching, learning and research.

Cristina López-Pérez has been a manager at the Open University of Catalonia (UOC) Library since 2006. She is responsible for Library Services for Learning. Previously, she was the manager of Documental Resources. She is a member of EDUCAUSE from UOC. She has a degree in Librarianship (University of Barcelona, 2000) and a technical specialist degree in Knowledge Management (UOC, 2004). She is now studying for a Bachelor's degree in Documentation at UOC.

Daniel McDonald is an alumnus of the University of Southern Queensland (USQ) and is currently a health librarian with the Toowoomba Clinical Library Service. He was recently shortlisted for the HLA/HCN Innovation Award and the ALIA/IOG Excellence Award. In June 2011 he spoke at the Australian Pain Society Annual Scientific Meeting about his work with the Acute Pain Service in Toowoomba.

Sophie McDonald is an Information Services Librarian at University of

Technology, Sydney (UTS) Library. She is involved in investigating the ways in which Web 2.0, mobile devices and emerging technologies can be used to develop new information literacy programmes and library services. Sophie has a keen interest in the future of information literacy and developing more dynamic programmes through experimentation and play. She sees information literacy as a part of ubiquitous learning in physical, digital and mobile spaces and works with social media sites for UTS Library as a means of engaging with clients in their online communities. As a blogger, Sophie is interested in the idea of the 'library of the future' and the role librarians play in shaping that future. As a reader, she particularly enjoys vampire fiction in hard copy and on her iPad.

Annika McGinley chairs the Library Emerging Technologies Advisory Committee at Southern Cross University (SCU) in Lismore, New South Wales, where she is the eReadings Coordinator. She has also worked as ePublications Librarian (Business and Law) for SCU's institutional repository and is currently programming the library's mobile website. Annika is pursuing a Master of Information Services degree at Edith Cowan University in Perth, focusing on best practice and benchmarking in library emerging technologies and, in particular, applying ITIL v3 (Information Technology Infrastructure Library v3) in an academic library setting.

Dr Damien Meere received his BSc degree in IT&T (Information Technology and Telecommunications) from the University of Limerick, Ireland, in 2005. He is currently completing his PhD within the Telecommunications Research Centre at University of Limerick. His research is focused on the provision of contextualized and personalized mobile e-learning (m-learning) services within a variety of informational environments, primarily facilitated through the use of intelligent mobile Agents. As well as developing blended learning environments, Damien's academic interests include computer networking. He is a Cisco Certified Network Associate (CCNA) and a Cisco Certified Academy Instructor (CCAI).

Keren Mills is the Digital Services Development Officer in the Open University Library. She is responsible for Library Services website content and usability, providing support and guidance to content authors to ensure quality and integrity of library web content. She is also involved in monitoring, evaluating and implementing mobile technology and audio, video and interactive software applications within digital library service innovation. In 2009 Keren undertook a ten-week research fellowship at Cambridge University Library, focusing on finding out how people use their mobile phones to access or use information on the move. The results have informed development of library services to mobile devices at both Cambridge and the Open University libraries. The findings were reported at the Second International M-Libraries conference in June 2009.

Sam Moffatt is currently a Systems Co-ordinator in the University of Southern Queensland (USQ) Library, working in areas from open repositories with e-prints to resource collection repositories and library catalogue search using VuFind. In this work he has worked on building alternate access to information to support mobile services. Sam also works on the Joomla! open source content management system and is doing work using Joomla! to provide mobile web services side by side with the desktop equivalent.

Niu Xianyun is the Intermediate Engineer for Information and Network Department, National Library of China.

Dr Máirtín Ó Droma, BE (NUI), PhD (NUI), CEng, FIET, SMIEEE, is Director of the Telecommunications Research Centre and Senior Lecturer, Electronic and Computer Engineering Department, University of Limerick (UL); Member, Governing Authority, UL; Subject Matter Expert, Institute of Electrical and Electronic Engineers (IEEE); Invited Expert, International Telecommunications Union (ITU), Geneva; Chairman, Royal Irish Academy's Communications and Radio Science Committee; President, Official Commission Members Ireland, International Union of Radio Science (URSI); Ireland's delegate to European Science Foundation (ESF) Cooperation in the field of Science and Technology (COST) Research Actions: 'Wireless networking for moving objects' WiNeMO (IC0906) and 'RF/microwave communication subsystems for emerging wireless technologies' RFCSET (IC0803); and UL's delegate to the national PRTLI-5 Telecommunications Graduate Initiative (TGI). His previous activities and posts include: founding partner and steering committee member of the European Union Framework Programme Networks of Excellence TARGET (Top Amplifier Research Groups in a European Team) and ANWIRE (Academic Network for Wireless Internet Research in Europe). Ireland's delegate to ESF COST Research Actions: 'Modelling and simulation tools for research in emerging multi-service telecommunications' (285) and 'Traffic and QoS management in wireless multimedia networks' (290); Lecturer, University College Dublin and National University of Ireland, Galway; and director of a number of engineering companies. His research interests include: behavioural modelling linearization and efficiency techniques in multimode, multiband, multicarrier broadband nonlinear RF (radio frequency) power amplifiers; complex wireless telecommunication systems simulation; wireless network and protocol infrastructural innovations and new paradigms, especially the Ubiquitous Consumer Wireless World (UCWW); and new e-learning and m-learning infrastructural paradigms.

Dr Mícheál Ó hAodha lectures (part time) at the University of Limerick, where he teaches on a number of HPSS (History, Politics and Social Studies) courses relating to the history of Irish migration. He has published more than 40 books, including: *Irish Travellers: representations and realities* (2006) and *The Turn of the Hand: a memoir from the Irish margins* (with M. Ward) (2009). Between 2006 and 2008 he was an AHRC (Arts and Humanities Research Council) scholar in the School of Arts, Histories and Cultures, University of Manchester. He has a particular interest in the use of technology as a means to circumvent barriers in educational access for groups that have traditionally been 'outside' the second-level and third-level sectors and has written many journal papers relating to mobile learning and more flexible educational access. His next book is due out shortly from Peter Lang, Oxford.

Helen Partridge is based in the Faculty of Science and Technology at the Queensland University of Technology (QUT). She has published widely in the area of teaching and learning and has won a number of teaching awards, including the 2004 and 2005 QUT Vice Chancellors Distinguished Teaching Award. Professor Partridge is the co-ordinator of QUT's library and information science education programme. In 2008 she received one of eight Teaching Fellowships from the Australian Learning and Teaching Council (ALTC). The fellowship established guiding principles and models of best practice for 'LIS education 2.0'. In 2009 Professor Partridge received $219,000 from the ALTC for a research project that will re-conceptualize library and information science education for the 21st century. Working with 11 other universities, she is leading a project aimed at establishing a framework for the education of the information professions in Australia. Professor Partridge's work in LIS education has recently been recognized through the receipt of a European Commission's Erasmus Mundus Scholarship. Through this scholarship she was a visiting scholar at the Oslo University College teaching in the International Master in Digital Library Learning. Professor Partridge is currently a visiting academic at the Oxford Internet Institute, Oxford University, where she is undertaking a study exploring the information practices of people in social media.

Ela Volatabu Qica has worked at the University of the South Pacific (USP) Library for 12 out of 18 years in her career. She currently works as Digitisation Librarian and has wide experience in library technical operations, serving both local and regional USP campuses. Her interest is in the area of information technology, in particular, its proactive use in the Pacific region to build on capacity so that cultural knowledge and oral tradition are preserved, accessed and shared. She holds Librarianship and Business Information Technology tertiary qualifications from Canberra University and the Royal Melbourne Institute of Technology University, Australia.

Jenny Raubenheimer is currently Director: Information Resources Distribution at the Library of the University of South Africa (Unisa). She is responsible for service delivery to remote clients who cannot visit the library in person. Jenny holds a Master's degree in Information Science from the University of South Africa; and also a Bachelor's degree in Afrikaans and Nederlands and the Higher Education Diploma, both from the University of Pretoria, South Africa. She has published articles on document delivery in South African and international journals and has presented papers within this field at various conferences worldwide.

Sarah-Jane Saravani is currently the manager of the Learning Hub at the Waikato Institute of Technology, Hamilton, New Zealand. She has responsibility for the provision and development of access to a range of information and learning resources, and for services to staff and students to support their development as lifelong learners in the 21st century. She is the current convenor of the ITPNZ Libraries Forum and the Forum's representative on the Strategic Advisory Committee to the National Library of New Zealand. She is involved with international standards committees on IT in education.

Hassan Sheikh is currently the Head of Systems Development team at the Open University (OU) Library in the UK and manages systems developers and IT support staff. He is the technical leader on several internal and external projects, including the continuous development of the OU Library websites, collaboration with several international institutions on the development of mobile library services, shaping the OU Library Systems Strategy and advising the OU Library senior management and staff on the latest technology trends in the world of libraries. Hassan has several years' experience in programming and usability evaluations and worked at the Open University Institute of Educational Technology for four and a half years before joining the OU Library in October 2006. His research interests include evaluating emerging (in particular library-related) technologies, development of mobile library services, programming, usability and user interface design, enabling interoperability between different systems and tracking the website users' behaviour.

Dr Stanimir Stojanov, PhD, ACM, IEEE, received his Bachelor's and Doctoral degrees from the Humboldt University of Berlin in 1978 and 1986, respectively. He is Lecturer and Head of the Computer Systems Department at the University of Plovdiv, Bulgaria. His research interests include: context-aware and adaptable software architectures, intelligent agents and multi-agent systems, agent- and service-based architectures, middleware, e-learning tools and environments and mobile e-learning services. Dr Stoyanov has served on the technical programme committees of many international conferences and workshops.

Margie Wallin is a Liaison Librarian at Southern Cross University, Coffs Harbour, New South Wales, and one of her dreams is to make the process of navigating the 'world of information' a little easier for students and staff. Providing access to information in a mobile setting is one way of achieving this goal – especially in terms of flexibility and access.

Andrew Walsh works at the University of Huddersfield as an academic librarian, where his main role involves subject liaison for the schools of Education and Professional Development and Music, Humanities and Media. Andrew is particularly interested in information literacy, the use of active learning within library sessions, the application of mobile technologies within the library environment and making use of appropriate Web 2.0 technologies. In recent years he has delivered conference papers, including keynotes, and published articles on information literacy, active learning and mobile learning in libraries; has co-written a book on active learning tips for librarians; and has written several book chapters on social media and mobile learning. Most of his recent publications and talks can be found via the University of Huddersfield repository (http://bit.ly/lilacAW). Andrew is an active researcher–practitioner, including studying part time for an information literacy-related PhD at the University of Huddersfield.

Wei Dawei has a Master of Public Administration degree and is an engineer. He graduated from Beijing Institute of Petrochemical Technology in July 1999. Since then, he has been working in the National Library of China (NLC). He has successively acted as Manager of the System Management and Maintenance Department, Beijing National Library Digital Technology Co., Ltd; Deputy Director, Acting Director and Director of NLC Automation Department (NLC New Technology Research Center); Director of NLC Computer and Network System Department (NLC Library Technology Research Center). He was appointed NLC Deputy Director in June 2009. He is currently the SC Member of the IFLA Section of Information Technology (2007–2011, 1st term).

Xie Qiang is Deputy Director of the Coordination and Operation Management Division, National Library of China.

Anna Zuñiga-Ruiz has been a manager at the Universitat Oberta de Catalunya Library since 2005. She is responsible for Library Services for Research. Previously, she was the manager of Library Services. She is Chair of the European Distance Teaching Universities (EADTU) Task Force on Libraries. She has degrees in Librarianship (University of Barcelona, 1993) and Sociology (Autonomous University of Barcelona, 1995), and has an MBA (part time, Pompeu Fabra University, 2005).

Foreword

Stephen Abram

I had the distinct privilege of being invited to provide the opening keynote for the Third International M-libraries Conference in Brisbane, Australia (11–13 May 2011). The conference theme, 'Mobile Technologies: Information on the Move', explored and shared work carried out in libraries around the world to deliver services and resources to users 'on the move' via a growing plethora of mobile and hand-held devices. Previous M-libraries Conferences (2007 Open University in the UK and in 2009 co-hosted by the University of British Columbia and Athabasca University) laid the foundations for international conversations about the challenge in libraries to meet the needs of the mobile user.

As the Conference acknowledged, we are seeing the emergence of a new user dynamic, as evidenced in the plethora of mobile devices as well as the app, smartphone and mobile culture that has started to pressure traditional and innovative institutional libraries to address and support the mobile user.

These proceedings share some of the presentations and research that was presented at the Third M-libraries Conference. My own keynote is available as a Slideshare and on my blog, Stephen's Lighthouse.

Over the years, we have learned that mobility has the power both to enhance the library experience and to disrupt. At this conference we explored how to tell the difference between an opportunity and a threat. What are the opportunities in learning, research, education, management, etc.? From my point of view there have been a few strategic shifts that are caused, or at least magnified, by the mobile revolution. And these changes are just in their nascent stages at this point. They include:

- the shift from desktop/laptop environment to one that is designed for a small screen and sound
- the shift from potentially shared network devices (desktops and laptops) to truly personal devices
- the shift to a sharing and collaboration ecosystem from one that is a constrained networked architecture but founded in personal social and professional networks with permeable boundaries
- the shift from enterprise-controlled architecture and machines to personally controlled and owned devices
- the shift to an information and communication ecology that is global rather than one that is bounded by institutional or national cultural norms and rules
- the shift to an end-user device that is fundamentally based on interpersonal and team communication and collaboration and supportive of social networking as its core, and outside the bounds of institutional controls
- the emergence of a world where discovery and learning is largely dominated by interdisciplinary and cross-disciplinary studies and one in which the collaboration of scientists and researchers goes well beyond the published record, but in advanced virtual communities of practice that transcend enterprise and institutional work teams.

This is an exciting time for libraries and library professionals. The software, applications and devices are finally moving into alignment with library values and mandates. Digital content is growing fast and some disciplines are approaching the ability to practise successfully with digital information the majority of the time, and sometimes with much better transformational ways of discovery and research. Libraries are social institutions on a global scale. Librarians are social animals. Our values support worldwide sharing, freedom of access to information and knowledge, openness, learning, discovery and research, and the curation and preservation of cultural properties. Mobile creates a newer platform that aligns with our goals.

Mobile sits at the confluence of people, content, community, curation, connection and contact. It sits in the centre of the sweet spot for the situations that librarians touch and have a positive impact upon: information, organization, discovery, preservation, culture, learning, research and more. As such, this conference and its predecessors play an important role in learning how to adapt to a new ecology. As I often say, the dinosaurs didn't become extinct because the climate changed. They disappeared because they didn't adapt fast enough to the change. Indeed, they were limited by brains the size of walnuts, so we have an advantage. With our bigger brains and our better communication abilities, we can adapt. Conferences like this help to show the way.

As we move towards discovery, research and learning ecologies that are dominated by electronic content, e-learning and support for multiple learning styles so that everyone can participate in more effective ways, we will see the impact of mobile technologies and applications rise to the forefront. As communication and data costs decline, as devices become increasingly affordable, and as human connectedness increases, we will live through this coming period where we can share and shape the changes as we see them emerging. We will also help to shape and create this new environment. It is now entering a period where the change will be exponential – and not as slow as in the past ten years. With conferences like M-libraries, we see the hard work being done to understand, experiment, innovate and adapt to these changes. As the old saw has it, 'The best way to predict the future is to invent it yourself.' If the quality of the presentations at M-libraries and in these proceedings is any indication, we can do it!

I hope you enjoy these proceedings, as a memory of your time at the conference, or as a record of what happened there in this library age, the Neocene of the mobile revolution. We have only two real choices: create the future or be a victim of it.

> Stephen Abram, MLS
> Vice President, Gale Cengage Learning

Introduction

Gill Needham

Since the publication of the second book in this series (Ally and Needham, 2010) the field of mobile service delivery has matured considerably and is transforming how libraries operate. Mobile devices have become ubiquitous and libraries around the world have begun to engage with the opportunities they offer – some in experimental mode and some as mainstream services. Advances in technology present both challenges and opportunities. The contents of this latest book reflect the variety of approaches being taken and present a rich panoply of experiences from many corners of the globe – 21 chapters from 11 countries as far apart as Spain and New Zealand, India and Japan, South Africa and the USA.

The book opens with a chapter by Mohamed Ally in which he describes the role of libraries in providing education for all. There are many global initiatives being implemented to provide at least a basic education to citizens around the world regardless of location, background and economic status. This chapter summarizes these education-for-all initiatives and explores how libraries could be transformed to provide education for all.

The remainder of the book is divided into three thematic parts. The first is Developing Mobile Services. It includes seven chapters from six countries, each describing the experience of researching, planning and implementing a diverse range of services for mobile users.

Deakin University in Melbourne, Australia, serves a large proportion of distance learning students as well as students based on campuses in three geographical locations. Colin Bates and Rebecca Carruthers explain in

Chapter 2 that these factors support the increasing use of mobile technologies. In their chapter they describe the development of mobile library services as part of a broader, university-wide mobile strategy. They have a programme of ongoing experimentation with new devices for both staff and students, and they share the feedback they have gathered from various small-scale trials.

The University of South Africa (Unisa) is one of the world's largest universities and, with more than 308,000 students, is the largest distance teaching institution in Africa. In Chapter 3, Jenny Raubenheimer points out that Africa is the fastest-growing mobile phone market in the world and that mobile phones are now regarded as the equivalent of personal computers for the African continent. While the majority of Unisa students have mobile phones, many do not have access to the internet via personal computers. Mobile delivery has therefore provided an ideal opportunity for Unisa Library to extend and enhance access to its services. Developments to date are described, as are plans for future innovation.

Vahideh Zarea Gavgani is an Iranian researcher whose work focuses on the potential of mobile technology for the delivery of consumer health information. In the study described in this chapter, she surveyed hospital patients receiving treatment for a range of conditions, to determine their preferred channels for receiving health information. Her results are presented in Chapter 4 and discussed in terms of their implications for service delivery. She explores different ways in which mobile phones might be used as a delivery channel to improve what she terms 'health literacy' in Iran and other developing countries.

The evolution of e-books and e-book readers is of crucial importance to the development of m-libraries. In Chapter 5 Anna Zuñiga-Ruiz and Cristina López-Pérez describe the implementation and evaluation of an e-reader loan service at the Universitat Oberta de Catalunya (UOC). Since the university's inception in 1995, UOC Library has offered its services online and has demonstrated a strong commitment to the innovative use of technology. The purpose of the e-reader loans service was to extend and enhance the service and support offered to students by providing e-readers on which they could access both their learning materials and library resources. The authors discuss the results of their evaluation, and implications for their future plans.

The University of the South Pacific is a complex institution, owned by the governments of 12 Pacific island countries and with 14 campuses. The

Library faces considerable challenges in supporting its diverse, highly distributed and multi-cultural student population. In Chapter 6, Ela Volatabu Qica discusses the opportunities offered by mobile technologies to strengthen the Library's role in providing services and support. She reviews the services offered by Fiji's major mobile telecommunications providers and the extent to which they will allow the Library to meet the needs of the student body.

While the majority of m-libraries' service developments have focused on text-based information, Margie Wallin, Kate Kelly and Annika McGinley at Southern Cross University (Australia) have taken a different approach. Chapter 7 describes a research project to investigate the potential of providing academic content in audio form, thus addressing different learning preferences and avoiding the problems of reading text on small screens. They report and discuss their findings, and implications for future development.

In Chapter 8 Daniel McDonald and Roger Hawcroft, from Toowoomba Clinical Library (Southern Queensland), describe the inception, development and progress of an innovative service to provide clinically orientated audio presentations to health professionals in their region. Using the opportunities provided by mobile technology (iPods in this case), they built on their experience of the successful use of audio to maximize access to quality medical information for busy (and frequently mobile) clinicians, to build an ambitious and highly successful suite of services.

Part 2 is entitled People and Skills. The six chapters focus on two aspects of skills development: the skills library staff need to develop in order to support mobile users; and mobile delivery of information literacy skills for student users.

Sarah-Jane Saravani and Gaby Haddow carried out an investigation of the preparedness of library staff in the vocational education sector in New Zealand and Australia to implement m-library services for their users. They surveyed librarians in 14 institutions about their competence in using mobile technologies and followed up with in-depth interviews. They discuss their findings in Chapter 9 and make recommendations for ways in which the resulting challenges could be addressed by both library management and professional education. A complementary chapter, Chapter 10 by Kate Davis and Helen Partridge at Queensland University of Technology, describes a research project which set out to identify the skills, knowledge and attitudes required by the m-librarian. The authors carried out in-depth interviews with librarians providing mobile services and used their analysis to propose ways in which the librarian's traditional skills and attitudes need to be enhanced in order to meet these evolving demands. Both chapters will be of particular interest to those involved in professional curriculum development.

Four chapters provide contrasting but related experience of using mobile technologies to deliver information literacy skills development. In Chapter 11 Sophie McDonald describes a range of innovative activities at the University of Technology, Sydney, all of which aim to use m-technologies to engage students in developing their skills. She and her colleagues have made extensive use of Quick Response (QR) codes for everything from a treasure hunt as part of a Library Fun Day for new students, to providing instructions for the use of a self-service machine. Other initiatives include the provision of mobile searching workshops and low-cost vodcasts uploaded to YouTube.

In a similar space, a group of librarians (Julie Cartwright et al.) at Charles Darwin University chose to focus specifically on QR codes in order to investigate their users' knowledge and use of mobile technologies. They developed and evaluated a highly sophisticated library treasure hunt activity using QR codes and a range of social tools which ran over a four-week period. They present and discuss their results in Chapter 12 and reflect on the experience overall.

Andrew Walsh and Peter Godwin are librarians at two UK universities (Huddersfield and Bedfordshire). Their chapter, Chapter 13, falls into two distinct sections. The first is an analysis of the concept of mobile information literacy, based on a review of relevant literature and their own local research work. They suggest that, because of the impact of mobile technology, the library community needs to rethink the information literacy needs of users. The second part of their chapter consists of a review of their experience as providers of mobile information literacy in their respective institutions. A range of innovative interventions are described, with evidence of take-up and impact.

At La Trobe University in central Victoria, Australia, the postgraduate students enrolled on a Diploma of Education programme were each given an iPod Touch to enable them to download their learning materials and share their work with others. In chapter 14 Iris Ambrose describes the way in which she and her library colleagues seized the opportunity to use these devices to introduce the students to online library services and searching skills. The sessions are described and the results of the subsequent evaluation presented and discussed, as are the implications for future practice.

Part 3 is entitled Focus on Technology. Chapter 15, by Wei Dawei, Xie Qiang and Niu Xianyun, describes the history and implementation of the National Library of China's suite of mobile services. These services have evolved since an initial launch of SMS services in 2007 and continue to develop to meet the needs of a growing population of mobile phone 'netizens' (303 million in 2010, with 859 mobile phone users overall). The authors present the range of developments they have achieved and discuss the drivers, constraints and current and future challenges.

Chapter 16, by Seema Chandhok and Parveen Babbar, gives a comprehensive overview of the phenomenal explosion of mobile phone usage in India (811 million subscribers in 2011) and details the ways in which the potential is being harnessed across the country for learning and libraries. They describe a diverse range of initiatives involving both public and private sector institutions and provide inspiration for others at earlier stages of progress.

On a smaller scale, in Chapter 17 Jim Hahn presents his rapid ethnographic study of users of mobile computing, carried out at the University of Illinois in the USA. He set out to explore the impact of context on use by observing and investigating the use of library iPads by a small group of students while riding a campus bus. He describes his methodology and presents the results and their implications for future work.

Hassan Sheikh and Keren Mills describe the background and development of mobile services at the Open University Library in the UK in Chapter 18. They focus on the development process itself, following user requirement gathering, the experience and lessons learned to date, future plans and some hints and recommendations which may be helpful to others.

Keiso Katsura identified a shortcoming in the ways in which library opening hours are displayed and published. He observed from earlier studies that users rate library opening hours as important information, but he recognized that there was no easy way of gathering up-to-date and accurate information about which libraries in an area are open at a particular time. In Chapter 19 he describes the system he has developed, based on CGI (Common Gateway Interface) technology, to make this information available dynamically to mobile users.

In Chapter 20 Damien Meere and his colleagues from the University of Limerick in Ireland and Plovdiv University in Bulgaria describe their research, which is developing and testing a new architecture for providing both mobility and a personalized information environment for library users. The architecture is based on an InfoStation design model whereby users can access a range of m-services through a distributed network of intelligent wireless access points.

The final chapter in this section, Chapter 21, differs from the rest in format and content. Based on his extensive practical experience in the library at the University of Southern Queensland, Sam Moffat provides a simple step-by-step guide to developing a basic mobile website. This very practical chapter could prove to be an invaluable resource for any library embarking on mobile service development with only limited technical resources.

Finally, Mohamed Ally rounds off the volume with his concluding thoughts.

References

Ally, M. and Needham, G. (2010) *M-Libraries 2: a virtual library in everyone's pocket*, Facet Publishing.

1

Education for all with mobile technology: the role of libraries

Mohamed Ally

Introduction

Recently, there have been many initiatives to provide basic education to everyone in the world, regardless of their location, economic status and background. Among these initiatives are Goal 2 of the United Nations Millennium Development Goals, to provide universal primary education to all children, and Article 26 of the Universal Declaration of Human Rights, which states:

1 Everyone has the right to education. Education shall be free, at least in the elementary and fundamental stages. Elementary education shall be compulsory. Technical and professional education shall be made generally available and higher education shall be equally accessible to all on the basis of merit.
2 Education shall be directed to the full development of the human personality and to the strengthening of respect for human rights and fundamental freedoms. It shall promote understanding, tolerance and friendship among all nations, racial or religious groups, and shall further the activities of the United Nations for the maintenance of peace.

Other initiatives, such as the University of the People, the Open Content Alliance, the World Digital Library, the Khan Academy, Open Education Resources, user-generated content, etc., are also being developed. These initiatives, with the help of educators and librarians and the increasing use of mobile technology by people the world over, will make education for all possible.

The Education for All movement recognizes that all inhabitants of the world must achieve at least a basic level of education in order to function in society. The United Nations has some specific goals for Education for All (UNESCO, n.d.), which include:

- Ensuring that by 2015 all children, particularly girls, children in difficult circumstances, and those belonging to ethnic minorities, have access to and complete, free and compulsory primary education of good quality.
- Ensuring that the learning needs of all young people and adults are met through equitable access to appropriate learning and life-skills programs.
- Improving all aspects of the quality of education and ensuring excellence of all so that recognized and measurable learning outcomes are achieved by all, especially in literacy, numeracy and essential life skills.

We are in the mobile generation. Mobile technology is being used by people to conduct everyday business and to accomplish everyday tasks. The technology is changing the way people work, learn, conduct business, interact with each other and access information. Mobile technologies can also be used to access education, especially by people living in developing countries where computers are less accessible. The use of mobile learning has been increasing over the last few years, rapidly in developing countries, which have the fastest rate of growth in the acquisition of mobile technology. In developing countries people are moving directly to wireless mobile technology. This provides an excellent opportunity for libraries to reach learners around the world, regardless of location. Education for All will provide the world's people with the knowledge and skills to function in society so that they are productive and have a decent quality of life. Education provides intellectual stimulus, which is needed by all, regardless of location, background or economic status. It raises people's self-esteem and enables them to see themselves as meaningful and contributors to society. Educational for All will prevent intellectual starvation, which is a major problem today because of the lack of education in certain segments of society. Why are the disadvantaged deprived of basic education when we have the means to reach people and provide it to them? This chapter describes the importance of providing education for all and the role libraries can play in doing so.

Initiatives to provide education for all

There are many developments that can make education for all possible. The challenge is how to provide easy access to information and learning materials from anywhere in the world.

Cloud computing: Cloud computing allows libraries to place information and learning materials on the web so that learners can access information from anywhere and whenever they need it. Libraries need to move away from building physical space, and instead to provide electronic access for learners so that they will be able to access information from anywhere and anytime, using information and communications technology. According to Jordan (2011), cloud computing allows for increased interoperation with all types of systems through shared applications programming interfaces (APIs) and massively aggregated data. This will allow libraries to reach more users and collaborate with other libraries so as to enrich the learner's access to information. The existence of the web means that cloud learning and learning materials can be accessible everywhere. Libraries will have to provide cloud services to learners, whereby learners will be able to access help and information from a variety of repositories, as required.

University of the People: This is the world's first 'tuition-free' online academic institution, dedicated to providing global access to higher education. The high-quality, low-cost global educational model is taking advantage of the growth of the internet and decreasing technology costs to bring university-level studies within the reach of millions of people around the world. The supporters and educators at the University of the People are facilitating a new wave in global education.

Open access press: An example of an open access press is Athabasca University Press (AU Press), which disseminates knowledge and research through open access digital journals and monographs and electronic media. The publications of AU Press are accessible to a broad readership through open access technologies, and thus contribute to open and distance education. This initiative is making a significant contribution to education for all.

Million Dollar Book Collection: The purpose of this initiative is to create a Universal Library that will foster creativity and free access to all human knowledge. Rather than having duplicate collections, libraries will be networked together to share collections, resulting in efficiency and cost savings.

Open Content Alliance (OCA): This alliance is a collaborative effort to build a permanent archive of multilingual digitized text and multimedia material. Its multilingual nature will enable it to cater for the requirements of people from different parts of the world. Its multimedia materials will cater for different learning styles. The capabilities of current mobile technologies support the use of multimedia, meeting the expectations of current and future generations of learners. Libraries will need to make the shift from textual information to multimedia.

Project Gutenberg: This project makes e-books freely available for people to download and read on mobile devices. As e-readers and tablet computers become more affordable, this initiative will become more important in the provision of education for all.

Online Books Page: This website allows people to access books that are available for free. It will encourage authors to make their books freely available on the web. The trend is to move to open education resources, which will help to achieve the goal of education for all.

World Digital Library (WDL): This global initiative will enable free access to information and learning materials. Libraries will be networked together to share resources and provide a seamless service to people around the world, thus avoiding duplication of resources and services to learners.

Khan Academy: The goal is to provide free, quality education to anyone in any location. The website currently has over 2100 videos and 100 self-paced exercises and assessments covering everything from arithmetic to physics, finance and history. This global initiative is replacing classroom instruction, and thereby making a contribution to the provision of education for all. Librarians will play a role by ensuring that individual learners get the right learning materials at the right time.

User-generated content: The use of wiki technology, as in Wikipedia, and social software is allowing users to generate their own content to be made available to anyone who wants to learn. Because information of this type is written by users who are interested in the topic, rather than by experts, the writing style is less academic than in formal sources and can be more suitable for use by peers. One problem with user-generated content is the accuracy of the information; however, comparison to information provided in 'reliable' sources has shown the incidence of errors to be similar (Rand, 2010). Librarians will have a major role to play in ensuring that learners have access to accurate and relevant information. A major benefit of the use of user-generated content, such as Wikipedia, is that learners have an opportunity to develop good writing and critical-thinking skills when creating content.

Open education resources: Educational institutions around the world are making their course materials available for free for educational purposes. This is enabling people to access quality learning materials at no cost and is making a significant contribution to education for all. Organizations such as UNESCO see the value of open education resources (OERs) and have created chairs in OER to promote the creation and use of OERs. Librarians will play a significant role in making OERs available to people around the world (Singh, 2008). Learning materials will need to be tagged and organized for easy access anytime and from anywhere.

Use of mobile technology to provide education for all

There is a technological revolution in the world, especially in developing countries, where access to education is greatly needed. In developed countries there is a shift from desktop to mobile technology, but in developing countries people are moving directly to mobile technology, rather than first acquiring desktop computers and then moving on to mobile technology. This is an excellent opportunity for libraries to provide services and access to learning materials to all people of the world, especially those in developing countries. We are in the mobile generation. Mobile web usage is expected to double within five years, overtaking the PC access to the web. At the end of 2010 there were an estimated 5.3 billion mobile cellular subscriptions worldwide (ITU, 2010). By 2015, 80% of the world's population will be accessing the internet on mobile devices (Johnson et al., 2011). Libraries must allow learners to access learning materials and services using mobile devices. Some libraries have already taken steps by digitizing information and making the library mobile friendly. The International M-libraries Conference brings together librarians, educators and mobile technology experts to share their experience and research on the use of mobile technology in libraries. The mobile innovations of libraries around the world are moving us closer to 'a library in everyone's pocket' (Ally and Needham, 2010). This is critical, since people the world over are already using mobiles to access services in other sectors. In the financial sector, people conduct bank transactions using mobile technology – 'in the pocket banking' (Economist, 2007). The travel industry is using mobile technology to make travel convenient for its customers, and people can access government information using their mobiles. The healthcare system is also using mobile technology to provide services to patients (Kenny et al., 2009; Taylor et al., 2010). People around the world are using mobile technology for entertainment ('entertainment in the pocket') and to shop ('in-the-pocket shopping'). People are also connecting with each other using mobile technology. Globally, the total number of SMS messages sent tripled between 2007 and 2010, from an estimated 1.8 trillion to an impressive 6.1 trillion (ITU, 2010).

Transforming libraries to provide education for all

Libraries around the world have already started to digitize information for access by computers and mobile technology. Also, libraries are increasing their rate of spending on electronic resources faster than their rate of expenditure on physical materials (Association of Research Libraries, 2008). At the same time, because of the economic downturn, libraries are examining ways to become more efficient while continuing to provide quality services to users (Powell, 2011; Stoffle and Cuillier, 2011). Digitization and the use of mobile technology can be viewed as opportunities for libraries to expand and to meet people's needs.

Libraries need to go one step further and to make education available to all by collaborating with some of the Education for All initiatives described earlier in this chapter. They also need to include mobile web access in their strategic planning, since mobile technology will replace desktop computers. Libraries should see these rapid technological advancements as opportunities for renewal, so as to provide better services to learners so that education for all can be achieved. According to Jordan (2011), although libraries are looking at technological innovations, many of their management processes remain based on technologies that were developed before the web. Libraries need to keep pace with the rapid conversion to electronic materials (text and multimedia) and the needs of the new generations of learners. These new generations of are very comfortable with mobile technology, which they are using in their everyday lives. As a result, they will demand that information, learning materials and library services be provided via mobile technology.

We are moving into the Library 3.0 world, where libraries will make use of Semantic Web technologies, provide ubiquitous service and access, personalize information for individual learners, provide virtual services and provide information specific to locations, resulting in location-based learning. Mobile and emerging computing technologies will result in ubiquitous access to information and learning materials, whereby people around the world will be able to access information from anywhere, at any time. Computing devices will exist everywhere, providing learners with seamless access to information. Libraries will need to provide ubiquitous services (Barnhart and Pierce, 2011) and access to information on multiple devices, which will be embedded in the environment – in appliances, vehicles, etc. The learning space is moving away from the classroom at a specific time to any place and any time.

Because of the information explosion, librarians have to ensure that learners have access to accurate information at the right time, especially in the world of user-generated information. Information and learning materials must be tailored to learners' needs. Libraries need to develop Library 3.0 systems to filter irrelevant information so that the right people can get accurate information when they need it. Further, when providing education for all, information needs to be tailored for different contexts and cultures.

Conclusion

In the electronic age, libraries have to change from a collection-centric to a user-centric mode (Giesecke, 2011). This is critical for libraries, since information will exist in the cloud, where they will have little control over who accesses the information, what information people access and when they access it. In Library 3.0, the computing system will control information globally and the role of librarians will be to provide services to users that support education for all.

In the *information age*, libraries provide services for people to access information. We are now in the *knowledge age*, and in the future will be moving into the *wisdom age* (Baltes and Kunzmann, 2003; Baltes and Staudinger, 2000; Taylor, 2011). What will be the role of the librarian in the knowledge age and the wisdom age? In the future there will be machine-to-machine communication, whereby databases will communicate information to each other so as to provide relevant information to users and create a knowledge base. What will be the role of libraries in machine-to-machine communication, where the machine will have the intelligence to provide the right information to the right person, at the right time, in the right place?

With the rapid expansion of distance education, open education, e-learning and mobile learning, learners will be able to access learning materials and information from anywhere and at any time. As a result, the role of the teacher will change dramatically in a 'teacherless' world. The teacher will become a tutor in the education system. In this paradigm shift, an important question concerns what the role of the librarian will be. Research is needed to discover the changing role of librarians in a 'teacherless' world. At the same time, training programmes for librarians need to change, so as to prepare librarians for the mobile world. Both present and future librarians need to change their mental models of how libraries function and how services are provided to users (Giesecke, 2011). With cutbacks in academic institutions, libraries have to reinvent themselves (Lowry, 2011). Use of mobile technology and the ubiquitous library will help to meet the challenges caused by cost cutting. According to Stoffle and Cuillier (2011), libraries are in uncharted territory. Now is the time for the library to design its future.

Because of the global availability of mobile technology, for the first time in history we have an opportunity to enable people the world over to access information and learning materials, so that we can have education for all. This is being facilitated by the many initiatives that are making learning materials available as OERs. The growing availability of OERs is making access to learning more affordable for anyone who wants to learn. We will know when the goal of Education for All has been achieved when all citizens of the world are able to attain to at least a basic level of education in order to function in society. Librarians have a major responsibility to help in achieving this goal.

References

Ally, M. and Needham, G. (2010) *M-Libraries 2: a virtual library in everyone's pocket*, Facet Publishing.

Association of Research Libraries (2008) *Electronic Resources vs. Total Materials Expenditures, 1993–2008: yearly increases in average expenditures*, ARL Statistics 2007–2008, 19.

Baltes, P. B. and Kunzmann, U. (2003) Wisdom, *The Psychologist*, 16 (3), 131–2.

Baltes, P. B. and Staudinger, U. M. (2000) Wisdom: a metaheuristic (pragmatic) to orchestrate mind and virtue toward excellence, *American Psychologist*, 55 (1), 122–36.

Barnhart, F. D. and Pierce, J. E. (2011) Becoming Mobile: reference in the ubiquitous library, *Journal of Library Administration*, 51 (3), 279–90.

Economist, The (2007) *Mobile Banking: a bank in every pocket*, (15 November), www.economist.com/opinion/displaystory.cfm?story_id=10133998.

Giesecke, J. (2011) Finding the Right Metaphor: restructuring, realigning, and repackaging today's research libraries, *Journal of Library Administration*, 51 (1), 54–65.

ITU (2010) *The World in 2010: ICT facts and figures*, www.itu.int/ITU-D/ict/material/FactsFigures2010.pdf.

Johnson, L., Smith, R., Willis, H., Levine, A. and Haywood, K. (2011) *The 2011 Horizon Report*, The New Media Consortium.

Jordan, J. (2011). Climbing out of the Box and into the Cloud: building web-scale for libraries, *Journal of Library Administration*, 51 (1), 3–17.

Kenny, R. F., Park, C., Van Neste-Kenny, J., Burton, P. A., and Meiers, J. (2009) Using Mobile Learning to Enhance the Quality of Nursing Practice Education. In M. Ally (ed.), *Mobile Learning: transforming the delivery of education and training*, Athabasca University Press.

Lowry, C. B. (2011) Year 2 of the 'Great Recession': surviving the present by building the future, *Journal of Library Administration*, 51 (1), 37–53.

Powell, A. (2011) Times of Crisis Accelerate Inevitable Change, *Journal of Library Administration*, 51 (1), 105–29.

Rand, A. D. (2010) Mediating at the Student–Wikipedia Intersection, *Journal of Library Administration*, 50 (7–8), 923–32.

Singh, N. L. (2008) The Librarian as Essential Key to Connecting Open Educational Resources and Information Literacy in the Academic World, *Open and Libraries Class Journal*, 1 (1), 1–9.

Stoffle, C. J. and Cuillier, C. (2011) From Surviving to Thriving, *Journal of Library Administration*, 51 (1), 130–55.

Taylor, J. D., Dearnley, C. A., Laxton, J. C., Coates, C. A., Treasure-Jones, T., Campbell, R. and Hall, I. (2010) Developing a Mobile Learning Solution for Health and Social Care Practice, *Distance Education*, 31 (2), 175–92.

Taylor, M. M. (2011) *Emergent Learning for Wisdom*, Palgrave Macmillan.

UNESCO (n.d.) *Education for All Goals*, www.unesco.org/new/en/education/themes/leading-the-international-agenda/education-for-all/efa-goals.

Part 1

Developing mobile services

2

Preparing for the mobile world: experimenting with changing technologies and applications for library services

Colin Bates and Rebecca Carruthers

Introduction

The field of mobile technologies is huge. This paper tries to limit discussion to the context of university libraries and the delivery of information resources. However, it is helpful to be informed by the wider educational environment and technology marketplace. We explore: the technology background and technology take-up by university library clients; how libraries have used and are using client-serving technologies; the Deakin University Library (the Library) e-resources profile; the Library's adoption of and experimentation with mobile-friendly applications, interfaces and technologies; and some outcomes and conclusions based on current experience and technology evaluations.

Understanding the client technology profile

Understanding library clients and their current use of information technologies is essential to providing services and resources to them effectively, while also enabling knowledgeable advice on mobile technologies and interfaces to be provided within a library context. There has been an exponential increase in the ownership of laptop or netbook PCs by university students and smartphone use has exploded amongst both the student and general population. A great number of undergraduate students have and use laptops, netbooks and smartphones in library spaces. This high take-up of mobile technology by students is evidenced by Australian research at Curtin University of Technology which showed that 'Three quarters (76%) had laptops or netbooks ... Approximately 99% of respondents had mobile phones: three-quarters of those (75%) were web enabled.' (Oliver and Nikoletatos, 2009, 723).

Data from the USA presented in the *ECAR Study of Undergraduate Students and Information Technology, 2010* (Smith and Caruso, 2010), shows the technologies that students own and how they use them, among a surveyed 36,000 undergraduate students from over 120 universities and colleges. This client-profile information for the USA and Australia largely pre-dates the release of the iPad.

The ECAR study (Smith and Caruso, 2010) provides information on the technologies the students were using. Over 80% of students owned a laptop, and over 10% a netbook (close to 89% saturation among undergraduate students), and over 60% of students owned an internet-capable hand-held device (Smith and Caruso, 2010, 9). These would mainly be smartphones and iPod Touch or PDA technology. As the ECAR data collection is 12 months old it is reasonable to expect a greater current take-up of mobile technologies than is indicated in the 2010 figures.

Building mobile library resources and services

We know from this client data that there is a very strong mobile technologies infrastructure among university students. Libraries have reacted to the mobile technology 'buzz' and moved to make it support both clients and provision of information resources. The huge investment by academic libraries in electronic resources pre-dates the more recent convenient mobile technologies. Many libraries also use other internet- or web-based applications, from e-mail to social networking, to offer and promote their services. Libraries can now look to expanding them to more mobile devices. Web-based information resources are of great benefit, particularly in terms of their ease and convenience of access. It is here that the greatest financial investments have been made by libraries.

Deakin University Library serves four geographically distant campus populations as well as a large cohort of distance education students around Australia and the world, so electronic resources are invaluable. The teaching needs of the University fit well with electronic and remote delivery of resources. Even better use can be made of these resources by their effective use and promotion in a mobile environment. Clients can then really exploit the possibilities of their mobile devices and, with the help of the library, go mobile in ways they have not yet thought of, given that the default use of many mobile devices is primarily social and for entertainment. This extends information options, and the reach of the library.

The Library has developed projects to make its resources and services environment ever more mobile. These tie in with wider moves by Deakin University towards providing mobile-optimized web resources for students using mobile devices (Figure 2.1).

Figure 2.1 Deakin web for mobile devices, http://apps.deakin.edu.au/m/ © Deakin University 2011

Key services are being 'mobile optimized' and will drive student demand and use on hand-held devices. The Library subject guides (http://deakin.libguides. com/) use the Springshare platform and are also mobile-device aware. Clients get added value through mobile access to services supporting study, research and clinical placements. There are very good reasons for libraries to promote mobile access options, and return on existing investments in online information resources is a fairly major aspect.

Deakin University Library and electronic resources

During the last few years the Library has seen a budgetary shift in terms of its purchase of print and electronic resources. This can be clearly seen in Figure 2.2 overleaf, where 2010 saw less than 30% of the Library's total spend go towards print resources, almost a reversal of the percentages for 2003. These figures are particularly interesting, considering the fact that the Library does not consider itself to be e-

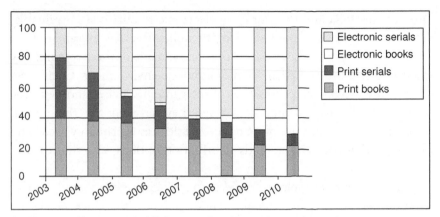

Figure 2.2 Percentage shift in spending from print to electronic resources, 2003–2010

preferred, but rather to be e-flexible (an electronic collection on top of a print base collection).

Serials has seen the fastest uptake and acceptance of electronic resources over print, reaching a plateau of around 85% of the total serial spend in 2006. This acceptance is due, in part, to an active programme of managed change, including: extensive marketing, one-to-one assistance and information sessions on new online resources. The Library continues to conduct these sessions, promoting resources and providing training so that its clients get the most out of resources.

This growth of e-resources, particularly the more recent growth in e-books, has prompted the Library to examine how its clients will be accessing and using them. Are they going to be able to extract the same flexibility and portability from these e-resources as from a printed book? This has prompted exploration of mobile technologies such as e-book reading devices, multi-function devices such as mobile phones, tablet PCs, along with netbook computers.

Kindle pilot

Deakin University Library is currently conducting a pilot study on lending e-readers and has selected Amazon's Kindle device for the study. The devices are available for two-week loan and contain a mixture of popular fiction and non-fiction titles. Clients are also able to request titles that are purchased and remotely loaded onto the device during the loan period. With every loan, users are encouraged to complete a survey. These have produced largely very positive responses. Interestingly there is no preference when it comes to the types of materials being made available on the devices. Since the advent of the Apple 'i' technology, users have come to expect certain things, including a colour touch screen; its lack of this feature has been a major criticism of the Kindle.

Survey responses also suggest that clients appreciate the ability to 'try before you buy', which this project allows. The Library, in conjunction with the IT Services Division, will expand this concept into a 'Tech Zone' trial, enabling users to test out a range of new devices within the Library.

Other technology on trial at Deakin

Deakin University is also internally trialling other devices, including the Kobo reader, the Sony eBook Reader Touch Edition, the Handii PC and the iPad. These devices were reviewed by Library staff and have been used in several 'play' sessions with local TAFE (technical and further education) and community library staff for group discussion and evaluation.

Because of the portability and flexibility of the iPad, Deakin University Library has allocated these devices to all team leaders, who have used iPads to reduce the amount of paper consumed in the office, making the move an eco-friendly one as well as a technological one. iPads and iPod Touches are also being used as mobile reference points by roving staff members, enabling client assistance 'on the fly'. These devices are also assisting staff in the collection of statistics online. With all new technologies, there are negatives associated with purchasing devices as soon as they become available. The industry is constantly evolving; you will always want something new, and as soon as you have it, it is out of date or obsolete. With the 'Tech Zones', there will be a higher overall usage of devices purchased for trial and assessment.

While looking at mobile technology one must remember that in the ECAR study (Smith and Caruso, 2010), 89% of undergraduate students owned a laptop or netbook computer – a multi-function device that is continually becoming smaller. Comfort and familiarity make for a happy technology user (most of the time), so it comes as no surprise that these are still in wide use, offering a familiar operating system and enterprise applications such as Microsoft Office. Unlike other portable devices, netbooks have USB ports, allowing for the easy sharing of documents.

Future devices

According to the feedback from our small-scale trials, future mobile devices must:

- not be locked down to a single e-book source (e.g. Amazon)
- be multi-functional and be able to run multiple programs at any one time
- be web enabled and have wireless functionality
- be able to handle all available e-book formats (e.g. ePub, PDF)
- have a colour touch screen
- be lightweight, while having a long battery life
- be able to run Flash web pages
- have the ability to add documents without needing to connect to a computer.

These requirements are often met in a netbook.

Lending technology?

The latest technology is tempting and can be expensive. Why not just lend it to people? This is not as simple a concept as it first appears. Deakin University Library has a long history of lending laptops to students, and with a range of support infrastructures this activity is successful. However, we know from the ECAR study (Smith and Caruso, 2010) that most students are buying their own laptops and smartphones. While lending laptops and Kindle e-book readers can be done, it must be done carefully and within the constraints of licence and usage agreements and local administrative requirements.

In particular, libraries are restricted in what applications and content may be offered on a device. The variety of licences, contracts and usage agreements all need to be abided by and are largely designed for individual use. This in fact means that while some content and mobile devices simply are not well suited for a lending environment, they can be very effective personal tools. Partnerships are possible with manufacturers and publishers but have to be carefully negotiated, and may not deliver all the preferred outcomes. The results of a range of partnership studies with publishers, Amazon and universities in the USA show there are mixed outcomes – for example, the Reed College Kindle Study (Marmarelli and Ringle, n.d.). There are likely to be pros and cons in each negotiated partnership situation. Usually, compromises have to be made, meaning that a partner doesn't quite get what it wants.

Before a programme of lending mobile devices is considered at the institutional level, a position should be developed in response to the following questions. Who will

- be responsible for maintenance and control of the devices?
- pay for applications, and 3G connectivity?
- administer authentication and subscription logins?
- manage content?
- provide training and instruction in use?
- repair and replace devices?

The list could become longer of course, as the detail is explored. Lending mobile devices is not a simple option.

If lending is difficult, an institution can, as an alternative, provide devices to staff or students. Recent examples of where students 'have been given free iPads' at several universities in Australia have been reported in the press (Maslen, 2011, 14). This could be seen as a populist approach, though undeniably done with the aim of improving educational outcomes. The *Chronicle of Higher Education* also reports some mixed outcomes for iPad use in the classroom (Kaya, 2010). The important answers to the question 'Why do it?' must lie in the productivity or educational outcomes to be achieved, and that requires some serious thought about and planning of institutional-level initiatives. By 'gifting' a device, the responsibility for using the device to achieve these outcomes passes to the owner.

Outcomes and conclusions

We have good information to work with, based on what we know of our clients, our resources and the currently available mobile technologies. This can inform the way in which a library service makes decisions on how to exploit the possibilities of mobile devices and electronic resources in order to give clients the flexibility and choices that can add value to their individual academic activities. Some key elements to consider are that:

- the extent of library investment in e-resources is key
- web-based delivery is platform independent and offers flexibility
- technology is constantly evolving
- client needs vary from one individual to another
- experimentation with and evaluation of devices by libraries is important
- it is important to understand what works for both clients and libraries.

These points, applied to particular library services and client cohorts, can help librarians to offer appropriate and sustainable client-centred services.

At the Deakin University Library there is growing practical experience of how useful an iPad or iPod Touch can be as a tool for mobile support of students and academics, whether within the Library or among the researchers and academics in laboratories and offices. This use of mobile devices for providing advice to clients was only made possible by having the devices in the Library for evaluation and testing in the first place.

Libraries should be investing in mobile technology for the knowledge and experience it provides, although not necessarily always with a view to lending the equipment. This activity helps to inform selection of library resources and the development of services and support to clients, and should be informed by the understanding that the final decision on any given mobile technology's usefulness lies with the individual who will be using it. To ensure the best value to clients, the mobile device needs to access the client-selected content, and the user must have control over the configuration that delivers the best personal results.

Our advice and experience as librarians who are informed and knowledgeable about mobile technologies and applications can help to map out effective services and resources to help our clients make the best decisions for their individual needs.

References

Kaya, T. (2010) Classroom iPad Programs Get Mixed Response, *Chronicle of Higher Education*, (20 September), http://chronicle.com/blogs/wiredcampus/classroom-ipad-programs-get-mixed-response/27046.

Marmarelli, T. and Ringle, M. (n.d.) *The Reed College Kindle Study*, http://web.reed.edu/cis/about/kindle_pilot/Reed_Kindle_report.pdf.

Maslen, G. (2011) Tablets Emerge as New Uni Tool, *The Age*, (8 March), www.theage.com.au/digital-life/tablets/tablets-emerge-as-new-uni-tool-20110307-1bl03.html.

Oliver, B. and Nikoletatos, P. (2009) Building Engaging Physical and Virtual Learning Spaces: a case study of a collaborative approach. In *Same Places, Different Spaces, Proceedings ASCILITE, Auckland 2009*, www.ascilite.org.au/conferences/auckland09/procs/oliver.pdf.

Smith, S. D. and Caruso, J. B. (2010) *The ECAR Study of Undergraduate Students and Information Technology, 2010 (Research Study 6)*, EDUCAUSE Center for Applied Research, www.educause.edu/ers1006.

3

Enhancing open distance learning library services with mobile technologies

Jenny Raubenheimer

Introduction

This chapter begins with a brief introduction to the University of South Africa in order to provide the background as to why the m-library development is so important for this institution.

The University of South Africa and its library

The University of South Africa (Unisa) is one of the world's top ten mega universities and the fifth largest ODL (open distance learning) institution. With more than 308,000 students registered already for the first semester of 2011, it is also the largest ODL institution in Africa. The Unisa student profile indicates significant diversity: of the students registered in 2011, 306,074 reside in Africa (including South Africa); 1135 in Europe, 790 in Asia, 359 in Oceania; 440 in South America and North America; and 2358 students are living with a disability (University of South Africa, 2011).

During the last decade Unisa has transformed itself into a truly ODL institution. Previously it conformed to the correspondence model of distance learning, with merely one-way communication to students. Study and library materials were predominantly in the printed medium and delivered by post. Requests to the Library likewise arrived via post or fax and depended on the printed medium.

The University's vision is to become 'the African University in the service of humanity' (University of South Africa, 2005). This drive to educate the continent in order to be of global service informs the importance of the ODL model to the realization of that vision. Unisa's ODL policy (University of South Africa, 2008) defines ODL as a multi-dimensional concept that, in practice, seeks to bridge the time, geographical, economic, social, educational and communication distances between students and the institution, between students and academics, between students and courseware, and between students and peers (University of South Africa, 2008). In support of this definition, the Library endeavours to bridge the distance between students and the Library; to use information technology, and mobile technology in particular, to promote open learning for remote learners; and, in light of its diverse student profile, to ensure equal access to information for all its clients.

The Unisa Library provides an important support service to the University. Its vision is 'towards *the* leading ODL Library in Africa' (Unisa Library, 2011). The Library has several branch libraries and access points at partner libraries throughout South Africa and in certain other African countries in order to support the many students who cannot visit the Main Library in Pretoria in person. In 2010, two mobile buses were added to supplement the Library's outreach services to remote students. Clients who cannot reach one of these libraries can use the centralized Request Services available in the Unisa Main Library in Pretoria to request information resources or literature reviews relevant to the teaching, learning and research activities of the University's staff and students.

In order to ensure that the Library's centralized Request Services remain relevant, new technologies are considered and implemented on an ongoing basis. The next section reflects on the evolution of this service during the last two decades.

Evolution of Request Services in the Unisa Library

The Unisa Library's Request Services was established when the University was South Africa's only correspondence university and, even then, it strove to support students who could not visit the Library in person. The Library has continually been compelled to consider mechanisms for faster and more accurate delivery to remote clients, as clients require Library material for course design, assignments and research, and they require it in a timely manner.

Over the last three decades, the evolution of the Library's Request Services has been shaped by the emergence of new technologies that have dramatically opened up client access to the information contained in libraries and elsewhere, and made the delivery of information resources faster and easier.

One of the major developments experienced by the Unisa Library took place in the early 1980s, when the card catalogue was automated; then, in the mid-1990s, the internet took the catalogue from being a database available only in the Library to one that is available worldwide.

Since the 1990s, an extensive list of subscription subject databases and an electronic reserve collection of recommended and prescribed reading for courses and assignments have been accessible to clients via personal computer, regardless of their geographic locations.

In order to complement the Library's enhanced access to information, delivery services were also enhanced, and the Library has deliberately shifted from the slower delivery of the postal system to the much faster delivery of the courier service to designated collection points. Naturally, while the Library remains sensitive to copyright restrictions, there has been a parallel shift from the postal delivery of photocopies of extracts from books or periodical articles to the online delivery of materials in electronic format as, for example, e-mail attachments.

The Library's Request Services were enhanced by the implementation of an online request mechanism via the Library Catalogue. This was well received by clients, but it was not particularly advantageous to the many students who had no access to personal computers or to the internet. They were unable to benefit from the electronic request and delivery services and still had to post their requests to the Library. In order to ensure equitable service delivery to all its clients, the Library therefore investigated an alternative solution to that of providing online access to personal computers.

The solution to the problem became apparent with the penetration of mobile technologies in South Africa during the first decade of the new millennium. Globally, during that period, the number of mobile phone subscribers rose to over four billion; by the end of 2008, statistics indicated that Africa was the fastest growing mobile phone market in the world and already had over half a billion mobile subscribers (Koutras, 2008). South Africa has the 17th highest number of mobile phone users in the world (Constantinescu, 2011) and more than 90% of

Unisa's students own a mobile phone (Arlow, 2011). These facts confirm that '[t]he library is at a crossroads of choices to make in responding to the . . . mobile society' (Friskney, 2009). It has been said that the mobile phone has become the personal computer of Africa (Macha, 2007) and, as such, it presented the Unisa Library with a solution to its problem. The Library started to investigate how the new mobile technologies could benefit library services to remote clients. In doing so it was further encouraged by the development of e-learning initiatives in Africa, as discussed at the e-Learning Africa Conference held in Zambia in 2010, as well as by the creation of the iBala Consortium in South Africa, which consists of information professionals who, encouraged by the capacity and potential of mobile technologies, are seeking to deliver information services in new, meaningful and effective ways. The members of the Consortium are seeking 'to build a community of information professionals who are passionate about mobile technology and mobile development in the library and information services field' (iBala Consortium, 2011).

The Unisa Library's shift towards m-library services

Investigations into mobile technologies revealed significant potential benefits in terms of access to Library services and information for the majority of Unisa Library clients. They would in particular benefit students and researchers 'on the go', as many of them spend so much time in transit. The following sections illustrate the benefits of the Unisa Library's integrated mobile technology for delivering services to remote clients.

Mobile access to the Unisa Library Catalogue

Clients have mobile access to the Unisa Library Catalogue with Innovative Interface's AirPAC product and can search for information. Mbambo-Thata (2009) reported on this development at an m-libraries conference. AirPAC's impact is currently being investigated, together with the introduction of new phones now in the marketplace. Unisa clients with smartphones can access the Unisa Library Catalogue at the following web address: http://m.oasis.unisa.ac.za, and clients with older web-enabled phones can use the classic interface at: http://millennium. unisa.ac.za/airpac.

Clients benefit from this service because a request for an item accessed via the mobile technology is captured for processing by the Library immediately it has been placed.

Mobile access to lending services

Library clients can view their loan records to ensure that they have not exceeded their quota before placing another request; view and cancel material on hold; and extend the loan period of borrowed material.

Mobile access to global information

In addition to making m-requests for research material housed in the Unisa Library's collections and accessible via the Unisa Library AirPAC, clients can also access the Online Computer Library Center's (OCLC's) WorldCat Mobile via their mobile phones, for example, and request material available from participating libraries worldwide to be delivered via interlibrary loan (OCLC, 2011).

Mobile access to the Reference Management Service

Mobile access is provided to the valuable mobile research tool called RefMobile, a product of RefWorks COS, which was launched in 2009 (RefWorks, 2011). This mobi-service significantly extends the Reference Management Service available to Library clients. They can access the most commonly used functions in RefWorks' web-based reference management system via web-enabled mobile phones, smartphones or personal data assistants (Kilmon, 2009). The Library also draws students' attention to the free reference management tools that are available, for example, Mendeley (for iPhone, iPad or iPad Touch) (Mendeley, 2011).

Mobile access to the Unisa Library's e-collections

The Unisa Library currently subscribes to approximately 23,000 e-books. These e-books can be read on many of the available e-readers. This has greatly enhanced the accessibility and usage of the Library's e-resources for clients who are on the go. The Library also informs clients of free e-books and specifically mobi books such as *Kontax*, the m-novel initiated by the Shuttleworth Foundation, which 'aims at developing the literacy skills of students while using their favourite toy', i.e. the mobile phone (Pambazuka Newa, 2009). With the introduction of e-book applications for the iPhone in 2008, e-books are now the third most popular category of downloads after games and entertainment (Temme, 2009).

As the Unisa Library subscribes to a vast number of subject databases, this e-content is increasingly available via mobile technologies such as mobile phones. This is particularly useful for searching for or reading Unisa Library full-text items while on the move.

Mobile access to the Unisa Library's training services

To support the Library's training initiatives, information literacy training programmes for all clients unable to visit the Unisa libraries are currently being developed for access from mobile phones or other mobile devices. The training content is designed to orientate and educate clients in the use of the Unisa Library. Camtasia software (TechSmith, 2011) is used to create podcasts in the appropriate formats.

Instant communication with Unisa Library clients

In light of the high number of Unisa students with mobile phones, communication to the mobile phones of remote students has become the norm for communication and service delivery.

A client receives an SMS acknowledgement upon receipt of a request, for example, and is provided with a reference number for tracking that request. When the requested material is dispatched, the client is again issued with a track and trace number for that specific transaction. The sending of library notices by SMS is currently being developed.

The Library had created a mobi site to link all the information to be communicated to Unisa clients via mobile phones using Zinadoo, a free mobile website creation tool (Zinadoo, 2008).

Another communication tool used for conveying information is the Quick Response (QR) code. These codes are useful for communicating information to mobile phones because they compress large amounts of information into a coded image. The Library uses these codes in the cover letters to students that accompany literature search results, linking them to important information such as instructional material, important URLs for accessing their course reserves, e-mail signatures, etc.

Further, the QR code expands the mobile phone user's access to information. For example, in 2010 the Library acquired a copy of Jules Verne's *Around the World in 80 Days*, which is enhanced with QR codes. The codes can be read by the client's mobile phone and will link the client to additional information about the places visited in the story, among other things. The book states that it is the 'Ubitour version guided by your mobile phone' (Verne, 2010). This marries print and electronic media in a way that is very exciting. The printed text can now possess many of the enhancements of online media.

Mobile service evaluation

The Library has benefitted from surveys conducted via mobile phone because clients receive and return the surveys immediately. The response rate is not only high, but also provides a valuable opportunity for clients' voices to be heard and for the Library to address reported issues without delay. The online survey software has proved to be excellent software for this purpose because it displays the survey perfectly on older types of mobile phone and this ensures the largest possible sample for mobile surveys.

Format changes for the use of mobile devices

In light of the Library's shift to m-services, the Library has introduced a text-formatting service aimed at clients living with visual or reading disabilities such as dyslexia. Material can be converted into the format of choice, e.g. MP3 or text (from print), with the help of the Plustek BookReader (Plustek, 2010) or Eye-Pal (ABiSee, 2011). Material can also be converted into the DAISY format (DAISY Consortium, 2011) by using the Dolphin EasyConverter (Dolphin Computer Access, 2011). The converted material can then be loaded onto a mobile device such as the BookSense or BookCourier, which are portable digital audio DAISY book players, or the ClassMate Reader, which has additional features for the dyslexic student. In support of the m-library, the Unisa Library provides these assistive devices on loan to students.

Another very useful application is CapturaTalk, which the Library often advises its clients to experiment with. CapturaTalk captures text from books, signs, leaflets, etc. using the phone's camera. It can recognize the text using optical character recognition and read it back to the client using high-quality text-to-speech voices (Mobispeech Partnership, 2011). This benefits clients with reading problems, as text can be read to them on their Windows mobile phone. It is ideal for clients who require literacy support for disabilities such as dyslexia, or for those learning English.

Conclusion

The future of m-libraries at the Unisa Library looks promising. The Library will continue to develop its services in this area, with a view to becoming a leader in m-library development. The work of the iBala Consortium will certainly help the Library to develop and standardize its m-library service delivery for the benefit of all library clients.

Acknowledgements

The author is grateful to Ms Sandra Hartzer, Deputy Director, Request and Information Search Services, Unisa Library, for input on the m-library services implemented at the Unisa Library, and Ms Karen Breckon, Information Search Librarian, Unisa Library, for technical advice and editing of the paper.

References

ABiSee Inc. (2011) *Eye-Pal: converts print into speech and Braille for the deaf-blind*, www.abisee.com/products/eye-pal.html.

Arlow, M. (2011) *Telephonic Communication Pertaining to Survey Results on Percentage of Unisa Students with Mobile Phones*, (14 June), unpublished.

Constantinescu, S. (2011) *Africa Subscribers to Mobile Phones*, Intomobile.

DAISY (Digital Accessible Information System) Consortium (2011) *Daisy Consortium home page*, www.daisy.org.

Dolphin Computer Access (2011) *EasyConverter: make information accessible for everyone home page*, www.yourdolphin.com/productdetail.asp?id=25.

Friskney, D. (2009) Talking with Doyle Friskney about how Mobility is Challenging Academic Libraries, *LibraryConnect Newsletter*, 7 (4), http://libraryconnect.elsevier.com/lcn/0704/lcn070403.html.

iBala Consortium (2011) iBala Consortium home page,

http://ibalaconsortium.pbworks.com/w/page/27489568/iBala-Consortium.

Kilmon, C. (2009) *RefWorks Launches Mobile Phone Interface: RefMobile users can now manage their research information on the go!* RefWorks, www.refworks.com/content/news/RefMobile_Press_Release.pdf.

Koutras, E. (2008) The Use of Mobile Phones by Generation Y Students at Two Universities in the City of Johannesburg, MCom thesis, University of South Africa, http://uir.unisa.ac.za/bitstream/handle/10500/717/dissertation.pdf.

Macha, N. (2007) Africa: mobile phones are Africa's PC, *Global Voices Blog*, (1 July), http://globalvoicesonline.org/2007/07/01/africa-mobile-phones-are-africas-pc.

Mbambo-Thata, B. (2009) The Library on the Phone: assessing the impact of m-phones at Unisa Library, *Second International M-libraries Conference, Vancouver, BC, Canada, 23–24 June*, http://ocs.sfu.ca/m-libraries/index.php/mlib/mlib2009/paper/view/36.

Mendeley (2011) *Mobile Devices*, www.mendeley.com/#features.

Mobispeech Partnership (2011) *A Revolutionary Way to Access Text*, CapturaTalk home page, www.capturatalk.com/index.asp.

OCLC (2011) *Introducing WorldCat Mobile*, www.worldcat.org/mobile/default.jsp.

Pambazuka News (2009) *South Africa: Kontax launched: m4Lit project*, (1 October), www.pambazuka.org/en/category/internet/59145.

Plustek (2010) *Plustek BookReader V100*, http://plustek.com/usa/products/audio-office-series/plustek-bookreader-v100/introduction.html.

RefWorks (2011) *RefMobile*, www.refworks-cos.com/refworks/RefMobile.

TechSmith (2011) *Camtasia Studio: screen recording and video editing software*, www.techsmith.com/camtasia.

Temme, P. (2009) Jumping into the Brave New World of Delivering Content to Mobile Devices, *LibraryConnect Newsletter*, 7 (4), http://libraryconnect.elsevier.com/lcn/0704/lcn070405.html.

Unisa Library (2011) *Vision Approved by the Library Executive Committee*, unpublished.

University of South Africa (2005) *2015 Strategic Plan*, Unisa Press.

University of South Africa (2008) *Open Distance Learning Policy*, http://cm.unisa.ac.za/contents/departments/tuition_policies/docs/OpenDistanceLearning_Council3Oct08.pdf.

University of South Africa (2011) *Headcount Enrolment for 2011*, available from Unisa's Institutional Information and Analysis Portal, http://heda.unisa.ac.za/indicatordashboard.

Verne, J. (2010) *Around the World in Eighty Days*, Unimark Books.

Zinadoo (2008) *Zinadoo: for all your mobile web needs*, www.zinadoo.com.

4

Use of mobile phones in the delivery of consumer health information

Vahideh Zarea Gavgani

Introduction

The mobile phone is increasingly becoming a significant healthcare enabler and delivery tool. It is evident that its popularity is due to its worldwide connectivity, coverage, ease of use, immediacy, cost-effectiveness and diverse capabilities. There are at least six times more mobile phone owners than computer owners in the world (Peres, 2009). Many predict that by 2013, 95% of the entire world population will have a mobile phone. The technology has become affordable and the infrastructure is widespread (Hyett, 2010). It is easier, cheaper and faster for physicians, residents and medical science students working in emergency rooms to exchange health information, consult their peers and seniors and keep up to date through mobile phones than via computer, internet and personal contact. Downloading, sharing copies of reference books, images, rare cases for discussion – all can be carried out effectively via mobile phone. Bluetooth is prevalent among medical science students and residents. Blackberries, PDAs and iPhones can be used as tools for connecting people socially and connecting them to the internet. In the patient–doctor relationship mobile phones can be used for the delivery healthcare and health information by either health information specialists or health providers. There are many studies reporting the use of mobile phones in the patient–doctor relationship (Prociow and Crowe, 2010; Varnfield et al., 2011), monitoring of patients with chronic diseases (Tirado, 2011), HIV education (Chiasson,

Hirshfield and Rietmeijer, 2010), behavioural research (Prociow and Crowe, 2010) and treatment and healthcare (Lim et al., 2011; Maddison et al., 2011; Riva et al., 2011).

Mobile phone take-up isn't a phenomenon of the developed world only. Developing countries, such as China, India and many countries in Africa, are adopting the mobile phone at an amazing rate (Bon Tempo, 2011).

On the one hand, patients seeking health information have better access to mobile phones than to computers; and on the other hand, it is evident that demand for evidence-based consumer health information (CHI) has increased. In consequence, it is essential to implement the capabilities of emerging technologies so as to connect evidence-based CHI, health providers and health consumers. In Iran, a developing country, the majority of patients require health information and information prescriptions for the treatment, diagnosis and management of diseases (Gavgani, 2010a). There are projects using mobile phones, especially SMS, for health information delivery to patients in Iran, although not library-based information services. According to personal observation and follow-up by the researcher, there is currently a programme for delivering AIDS awareness, control and healthcare education by the provincial health sector in Iran (Tabriz City, province of East Azararbayjan) [a description and the results of this programme have not yet been published]. A toll-free number has been inserted into electricity bills, encouraging the public to seek AIDS-related information by telephone consultation and by SMS. Because socio-economic factors affect people's healthcare, care-seeking and care-giving behaviour, in a country such as Iran AIDS-affected patients rarely search for healthcare and health information in person, and they usually conceal their health problem. In such a situation SMS, mobile phone consultation, even live chat via computer, are the best ways to obtain knowledge and care without being identified. Our priority now is to discover patients' preferred tools and channels for the delivery of health information, healthcare education and information prescriptions.

Objectives

This study aims to determine patients' preferences for receiving health information and information prescriptions via mobile phone and traditional, paper-based information services.

It asks the following questions:

- Through which channels do patients choose to receive information prescriptions or health information?
- From where do they seek health information?
- Is there any relationship between the age of patients and their tendency to use mobile phones and SMS in health information delivery?
- Is there any relationship between literacy rates and the inclination towards use of SMS and mobile phones?

Methods

A descriptive survey method with an open interview and structured questionnaire was used to gather data from patients in teaching hospitals and clinics in major teaching hospitals in Tabriz, Iran. At the end of each section an open-ended questionnaire was designed to gather the opinions of patients and record their suggestions. It is felt that this user driven-information could be very useful for policy making and in the design of a user-centred consumer health information system. Data was analysed using the Statistical Package of Social Science software (SPSS V.15) and was presented as simple frequencies. A chi-square test or Fisher's exact test was used to compare variables and a p-value of less than 5% was considered statistically significant.

Findings

A total of 139 patients with a variety of health problems (cardiovascular diseases and cardiac surgery, dermatology including burning and skin disease, internal medicine, ENT, arthritis, digestive and gastric diseases) were surveyed using a structured questionnaire and face-to-face interviews. The average age of participants was 40.04 ± 13.28 years. The youngest was aged 19 and the oldest was 80. The majority of patients, 116 (83.5%), were literate and 23 (16.5%) were illiterate.

Existing and emerging technologies for delivering consumer health information

Questions were asked about both existing and emerging technologies for delivering health information. Of all the methods for receiving health information – face to face, through print, by mobile phone, through mass media or through the internet – the highest percentage of patients preferred to receive their health information/information prescription via face-to-face contact (81%); next to that, mobile phone delivery was preferred by 75%. The lowest percentage of patients stated the internet was their preferred way to receive health information (Figure 4.1).

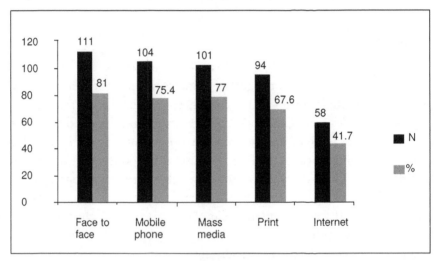

Figure 4.1 Existing and emerging technologies for delivering health information

Mobile phone capacities for delivering health information

The growing capabilities of the mobile phone have made it a popular device for exchanging health information.

Blackberries, PDAs and iPhones can be used to connect to people and to the internet. In the patient–doctor relationship mobile phones can be used to deliver both healthcare and health information, either by health information specialists or by healthcare providers. But the question is the patient's preference and convenience in relation to the mobile phone and its various tools for health information delivery. The mobile phone's capacities for delivering health information, such as voice message, SMS, data (text), Multimedia Messaging Service (MMS) for audio-visual files and images, and Bluetooth technology, were investigated (Figure 4.2).

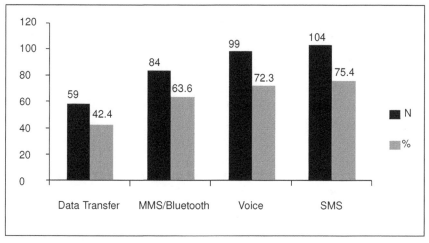

Figure 4.2 Patients' most preferred mobile-based tools for receiving health information

The majority of patients surveyed (104 or 75.4 %) indicated SMS to be the most preferred and convenient tool for receiving health information. After SMS came voice (phone call) (99 or 72%) and MMS (audiovisual file transfer) (84 or 63%). The lowest preference was for data transfer via mobile phone (42.4%) (Figure 4.2).

Access to a mobile phone and a wireless connection is a privilege among people in most nations around the globe. The world wide web (WWW), despite its potential for delivering and disseminating mass amounts of information in various formats, has not gained as much popularity as has the mobile phone. The reasons given by patients for their preference for the mobile phone are: portability of a mobile device, ease of use, impartial nature of advice, and a sense of personalization and privacy while receiving information through SMS. They stressed that information regarding one's health decisions was more confidential when received via a mobile device. Conversely, patients stressed the limitations of the mobile phone, such as limited memory capacity and limited display area, which make mobile phones inappropriate for reading long, text-based items of information. Patients' individual opinions about the WWW, collected via open-ended questions, indicate that the WWW is not preferred because of its high expense in comparison with the mobile phone, dependence on other devices and tools, low-speed connections, difficulty in application and time-consuming search processes.

The study reveals that giving information face to face, either orally or in any format such as a piece of paper or a URL written on paper, or via an SMS that addresses some recommendation, reminder or instruction directly to the patient or their care giver, is the most preferred way of receiving information prescriptions. This can be regarded as similar to a drug prescription that a patient receives from the healthcare provider and picks up from the pharmacy.

Association between literacy and preferred format of information delivery

The study found a statistically significant relationship between the literacy of surveyed patients and their preferred technologies for the delivery of information in print/paper-based formats (transferring information to mobile phone (P<0.001), internet (P=0.011)). It can be inferred that, as far as text-based formats and the use of the latest ICT facilities are concerned, the level of literacy has an impact on the health information prescription and information delivery in developing countries, specifically in countries similar to Iran. The study also reveals that there is no significant relationship between the literacy of patients and their preferred access channels. Literate, semi-literate and illiterate patients all prefer to receive information face to face (Table 4.1).

The study indicates that there is a statistically significant relationship between the age of patients and their preferred formats for receiving

Table 4.1 Relation between age of patients and preferred format of information

		Literate	Not literate	P-value
Print format	Yes	88 (75.9%)	28 (24.1%)	<0.001
	No	6 (26.1%)	17 (73.9%)	
Face to face	Yes	94 (82.5%)	20 (17.5%)	0.383
	No	17 (73.9%)	6 (26.1%)	
Mass media	Yes	82 (75.9%)	26 (24.1%)	0.593
	No	19 (82.6%)	4 (17.4%)	
Mobile phone	Yes	80 (70.2%)	34 (29.8%)	0.309
	No	19 (82.6%)	4 (17.4%)	
Internet	Yes	54 (46.6%)	62 (53.4%)	0.011
	No	4 (17.4%)	19 (82.6%)	

health information: specifically, SMS (P=0.001), phone call (P=0.001), WWW (P=020) and transferring information to mobile phone (P=0.014). Younger patients prefer to receive health information via SMS, mobile phone (call and data transfer) and WWW.

Discussion

The study reveals that patients in Iran, a developing country, are very interested in receiving health information and information prescriptions via mobile phone, followed by in-person and face-to-face delivery. In this regard, the results are similar to those of a survey in India, in which the highest percentage of patients (46.8%) were interested in receiving health information in person and 36.2% via SMS (Vishwa and Gavgani, 2009). Most patients prefer to receive health information via SMS, but fewer referred to transferring information/data to their mobile phone. Also they found Bluetooth technology more convenient than MMS. First of all, they were concerned about receiving large amounts of information via mobile phone because of memory and display problems. The patients' replies about mobile phone apps for the transfer/delivery of information indicate that in Iran (as in most developing countries), because of their low income, ordinary people use cheap and low-memory mobile phones rather than smartphones, Blackberries, PDAs or iPhones. Use of MMS service is limited to pay-per-service activation by two big mobile network providers, IranCell (www.irancell.ir) and Hamrah-e-Aval (www.mci.ir). This study also found that Iranian patients have a greater preference for SMS and mobile phone information delivery than for WWW, which tallies with the findings of the TGI Survey (TGI, 2009) and surveys in other developing countries (www.Frontlinesms.com). This preference can be compared with practices in developed countries, where WWW, discussion forums, e-mail and social networks such as Facebook, Twitter and blogs are widely used by health providers and patients to exchange information and patients' stories. However, there are limitations on the use of such Web 2.0 social networking tools in Iran.

However, paper-based formats still have a supportive role in learning processes; and audiovisual files like MP3, podcast, MMS and video files are still used by many patients. There is a need in developing countries, and specifically in Iran, for consumer health information (CHI) and information therapy to be provided in both ICT-based and paper-based formats. This is fully supported by the findings of the study (see Figure 4.1).

Conclusion

It can be concluded that patients prefer to use mobile phones with SMS, MMS and Bluetooth to exchange health-related information. The low speed of internet connections, dependency on other information infrastructures, the monitoring of social networks and the high cost of using the internet and related technologies are all obstacles or barriers in Iran and developing nations to receiving or exchanging health information via the internet. While patients consult friends and family members when they face health problems or are seeking health information (Gavgani, 2010a), health literacy needs to be improved in developing countries, especially in Iran, through the dissemination of correct, evidence-based CHI. Taking into account all the above-mentioned limitations, the best solution for such countries is to deliver information therapy, health information and information prescriptions to clients' mobile phones. According to the proposed model (Gavgani, 2010b), health information and information prescriptions can be easily ordered from librarians by a physician acting on behalf of a patient, via SMS; the evidence-based information or instruction can then be sent to the patient and/or physician via SMS, MMS for remote access, Bluetooth or Symbian, Android, and so on, for confidential access.

References

Bon Tempo, J. (2011) *Can Mobile Phones really Improve Health in Developing Countries? What the aid and development communities need {mHealth}*, iMedicalApps Team, (11 January),
www.imedicalapps.com/2011/01/can-mobile-phones-really-improve-health-in-developing-countries-what-the-aid-and-development-communities-need.

Chiasson, M. A., Hirshfield, S. and Rietmeijer, C. (2010) HIV Prevention and Care in the Digital Age, *Journal of Acquired Immune Deficiency Syndromes*, 15 (55), Suppl. 2, S94–7.

Gavgani, V. Z. (2010a) Health Information Need and Seeking Behavior of Patients in Developing Countries' Context: an Iranian experience. In Veinot, T. (ed.),

Proceedings of the 1st ACM International Health Informatics Symposium (IHI '10) held on 11–12 November in Arlington, VA, USA, ACM.

Gavgani, V. Z. (2010b) *Service Model for Information Therapy Services Delivered to Mobiles,* Facet Publishing.

Hyett, C. (2010) *Mobile Health in Developing Countries,* MedpageTodays', www.kevinmd.com/blog/2010/10/mobile-health-developing-countries.html.

Lim, S., Kim, S. Y., Kim, J. I., Kwon, M. K., Min, S. J., Yoo, S. Y., Kang, S. M., Kim, H. I., Jung, H. S., Park, K. S., Ryu, J. O., Shin, H. and Jang, H. C. (2011) Survey on Ubiquitous Healthcare Service Demand among Diabetic Patients, *Diabetes & Metabolism Journal,* 35 (1), 50–7.

Maddison, R., Whittaker, R., Stewart, R., Kerr, A., Jiang, Y., Kira, G., Carter, K. H. and Pfaeffli, L. (2011) HEART: heart exercise and remote technologies: a randomized controlled trial study protocol, *BMC Cardiovascular Disorders,* 11 (1), 26.

Peres, S. (2009) *Mobile Phones to Serve as Doctors in Developing Countries,* ReadWriteWeb, (20 February), www.readwriteweb.com/archives/mobile_phones_to_serve_as_doctors_in_developing_countries.php.

Prociow, P. A. and Crowe J. A. (2010) Development of Mobile Psychiatry for Bipolar Disorder Patients. In *Conference Proceedings of IEEE – Engineering in Medicine and Biology Society 2010,* 5484–7.

Riva, G., Wiederhold, B. K., Mantovani, F. and Gaggioli A. (2011) Interreality: the experiential use of technology in the treatment of obesity, *Clinical Practice & Epidemiology in Mental Health,* 7, 51–61.

TGI (2009) *Mobile Phone Market Booms in Iran,* KMR Update 19, http://globaltgi.com/news/documents/KMRUpdate19.pdf#page=2.

Tirado, M. (2011) Role of Mobile Health in the Care of Culturally and Linguistically Diverse US Populations, *Perspectives in Health Information Management,* 1 (8), 1e.

Varnfield, M., Karunanithi, M. K., Särelä, A., Garcia, E., Fairfull, A., Oldenburg, B. F. and Walters, D. L. (2011) Uptake of a Technology-assisted Home-care Cardiac Rehabilitation Program, *The Medical Journal of Australia,* 194 (4), S15–9.

Vishwa, M. and Gavgani, Z. V. (2009) Informing Clients through Information Communication Technology in Health Care Systems, *Issues in Informing Science and Information Technology,* 6, 585–93.

5

Deploying an e-reader loan service at an online university

Anna Zuñiga-Ruiz and Cristina López-Pérez

Introduction

The Universitat Oberta de Catalunya (UOC) is an online university born of the knowledge society and with a mission to provide people with lifelong learning and education. It was founded in 1995. The UOC offers student-centred learning with the benefits of personalized study, flexibility, accessibility and collaboration. In its online study programme, the UOC places the student at the centre of the learning process and provides him/her with the necessary distance learning resources for interaction with the whole of the university community. Learning at the UOC is oriented to responding to the needs of the student and takes into account the demands of the professional environment and technological and social developments.

UOC's educational model

The UOC's educational model is the university's main feature, which has distinguished it since its inception. It was created with the intention of responding appropriately to the educational needs of people committed to lifelong learning, and to make maximum use of the potential offered by the web to perform educational activity.

The learning activity is at the centre of the educational model (Gros, 2009). The students have three main elements with which to carry it out: (1) the resources, which include the content, spaces and tools necessary to performing the learning activities and their assessment; (2)

collaboration, a set of communicative and participative dynamics that promote the building of knowledge among classmates and teachers, through teamwork, in order to solve problems and develop projects; and (3) accompaniment, the set of actions performed by teaching staff to monitor and support students.

UOC Library: a virtual library

The UOC Library has offered its services online since its creation in 1995 (March Mir et al., 2010). It has displayed a strong commitment to new technologies throughout this time. It is important to note that the UOC virtual library model implies that traditional library services (loan, reference, document supply, interlibrary loan, information competences, training) are offered at a distance.

The UOC Library also offers remote access from the virtual campus to its electronic resources, without the need for additional configuration on top of the campus access. Since its creation, the UOC Library has provided its services and content description in three languages: Catalan, Spanish and English.

The Library contributes to the educational model through the classroom resources. The classroom resources document management service is the key component of the library's commitment to this educational model. The classroom resources are managed by the lecturers and library managers together (Cervera, 2010a).

The project TRIA! (Choose! in Catalan), developed by the UOC Office of Learning Technologies, aims to provide all UOC learning materials and resources in multiple formats, leaving it up to the student to choose their preferred format (MP3 and DAISY zip files, text and MP4, mobipocket, ePub and PDF A6). In 2010 most of the UOC learning materials were available in multiple formats.

Loan of e-reader devices and in-situ consultation service

During a three-month period from October to December 2009, the UOC Library piloted an e-reader loan and consultation service (Cervera, 2010b). The e-reader loan service was set up in line with the UOC's commitment to promoting the use of e-readers and e-books as an extension of the UOC 5.0 virtual campus in support of student learning. The objective of the project was to contribute to learning and teaching activities by providing e-reader devices for access to UOC learning materials in e-book and PDF formats, as well as for access to e-book content subscribed to by the UOC Library.

The e-reader loan service was piloted using 15 iRex iLiad e-readers, which were borrowed 37 times. By the end of the pilot there were 300 users on the waiting list. The Library e-book website included a presentation on the service, information regarding the e-book collections available, shortcuts to the UOC Library online public access catalogue (OPAC), the UOC learning materials search widget and the Library e-resources search widget. The results of the pilot encouraged the Library to consolidate the e-reader in-situ and loan service.

During 2010 the e-reader in-situ and loan service increased the number of e-readers from 15 to 35, with the acquisition of 20 Kindle 2 devices. At the beginning of 2011 the number of available e-readers was increased to 235 with the acquisition of 200 Lector e-readers. There are 5800 e-book titles available from most relevant publishers, and a large proportion of the UOC learning materials.

The e-reader loan service works the same way as for book loan. E-readers are an item within the UOC Library OPAC. Library users must choose the pick-up location, which can be either a UOC regional centre or the University headquarters. The e-reader can also be delivered to the student's residence. Once the e-reader has arrived at the chosen pick-up location the user is informed by e-mail and then has four days in which to pick it up. Lending the e-readers via the OPAC allows for the management of loans, reservations, users and of returned devices, which are checked over to ensure that they are functioning correctly and that documents uploaded by previous users have been deleted.

During the pilot project questionnaires were sent to e-reader users; on the basis of their answers and of the Library's analysis of the pilot project in terms of its internal procedures a number of problems were detected that the Library needed to address in order to be able to consolidate the e-reader loan service.

What we learned from the e-reader in-situ and loan service pilot was the following:

- 90% of the users rated the e-reader loan service positively or very positively.

- The service needed to increase the loan time and number of devices.
- 52% of users believed that content was not adapted to e-readers or could not be loaded onto the e-reader device.
- E-readers were used much more to read UOC learning materials and PDF documents downloaded from magazines subscribed to by the UOC Library than to read the e-book collection.
- There was an insufficient selection of e-books in Catalan and Spanish.
- Most users found it easy to load content onto the e-readers.
- On-screen reading was comfortable for most users.
- Better user guides needed to be developed by the UOC Library.
- Some users were using the e-reader service as a part of their e-reader shopping process.

As was already mentioned, the pilot results encouraged the UOC Library to consolidate the service. The following improvements were made:

- An e-reader model was chosen that was more user friendly for UOC users.
- The number of devices was increased.
- Bibliographic records were enhanced by including each subject title to increase access.
- Electronic versions of UOC learning materials were included in the OPAC.

Reasons for deploying an e-reader loan service

These are principal reasons, from the UOC Library's point of view, for deploying an e-reader loan service:

1 To offer library users access to the many kinds of learning materials available at the UOC through a single device that facilitates student study and learning.
2 The e-reader loan service fulfils users' needs for access to electronic learning content and materials.
3 To enhance access to and use of the e-book collections subscribed to by the UOC Library. The UOC Library invests a signification portion of its budget in subscriptions to electronic resources. Also, as a principle of its collection development policy, the UOC Library

has chosen to buy electronic content as opposed to print formats whenever possible. This allows the Library to provide better services to its users in terms of timeliness and content availability.

4 To align with the UOC's strategy on the development of learning materials in multiple formats and to promote e-devices as an extension of the UOC Virtual Campus. The e-reader loan service enhances and promotes this expansion because it offers students the option of experience with the ePub or PDF formats on different devices, anytime and anywhere.

5 To increase UOC Library's visibility within and beyond the institution.

6 The service also has the strategic objective of increasing the Library's visibility. Internally, it has allowed the Library to demonstrate its commitment to learning and teaching, as well as to connecting with users who have the desire and curiosity to experiment with new devices. Externally, in collaboration with the Communication Office, both the pilot and the service developed subsequently have allowed both the UOC and the Library to obtain exposure in the media.

Key issues for the deployment of e-books and e-readers

E-books and e-readers raise some complex issues for libraries, but they also affect content creators, as well as the business models of publishers, editors and e-reader manufacturers. For this reason it is not the authors' intention to provide an in-depth analysis of those issues, but to raise awareness about the difficulties libraries face.

The e-readers

The e-reader market is at present characterized by instability. In a short period of time a wide range of e-book reading devices (from notebooks to mobile phones, together with tablets, the iPad, the Kindle and other e-readers) have appeared on the market. Each type of device offers different functionalities, and they face a rather high rate of obsolescence. In this scenario it is not easy for a library to choose what to buy, but the Horizon Report (Johnson et al., 2011) does state that the e-reader is the next device to be adopted by learning institutions.

The acquisition of e-readers for the implementation of an e-reader loan service involves a substantial investment. The option chosen by the UOC Library was to buy the greatest number of devices possible by selecting a device that fulfils e-reader users' needs but is not the trendiest device available.

The e-books

In regard to e-books, from the UOC Library's point of view, there are three main issues that are still unresolved and that have a significant impact on libraries: (1) the model for purchasing e-books and the technical infrastructure needed to make e-books available from the UOC Library website, (2) what users can do with e-titles, and (3) the pricing or licence model for e-books.

In regard to the first point, there is no single purchasing model for e-book titles, and this fact does nothing to simplify an already complex issue. On the second point, most e-book titles can be included in the OPAC, but what users can do with them is diverse. In most instances it is necessary to link to the e-book platform from the OPAC; in other instances, the library can provide access to the content directly from the OPAC. While printing e-books is mostly possible, in some cases users cannot download the titles and load them onto their e-reader devices; in others it is possible to download the books only by chapters, or to download only a part of the book. The third problem that affects libraries is the pricing or licence model for e-book titles. Some publishers are reproducing the same licence model that was established when electronic journals first appeared; some are offering packages of titles or collections and a licence agreement that needs to be renewed periodically, while in other cases titles are bought in perpetuity. Other publishers are offering both packages of titles and the option of choosing single titles; in this case, however, there is a significant price difference. In most cases, buying single titles is far more expensive than buying a package. In yet other cases, publishers are selling only to individuals and not to libraries or institutions, or they are following the same model as for paper books, whereby the Library has to buy a number of individual items for student access.

The scenario described presents the UOC Library with important challenges because it is unclear whether the Library will be able to meet the licence and pricing models offered at the moment. It is also difficult for the UOC Library to increase e-book usage when all users have in mind paper books and the traditional loan service, which are far simpler than e-reader and e-book use.

The library's role in promoting new reading formats to increase the use of e-collections.

Libraries have always played an important role in promoting access to information, culture and knowledge – which is part of most libraries' mission. E-readers allow access to a huge amount of content, regardless of space and time. The UOC library has set up the e-reader loan service not only so as to increase the use of its e-collections but also because UOC learning materials are available for use on e-reader devices and this allows the UOC library to contribute to the excellence of the UOC educational model.

Conclusions

Whether the e-reader loan service in libraries will become a permanent library service is very difficult to say. It depends on many factors, among them, the e-reader or similar devices themselves, rate of market penetration, the title offer and prices, and the degree of simplicity involved in the process of buying an e-book and loading it into the e-reader. In any case, from now on e-book projects will focus more on content than on the e-reader devices, because content is what users value.

With regard to university libraries, it is important for acquisition policies that prioritize e-books in relation to paper to be supported by the university's board of directors. In the case of UOC Library, this support was received in May 2011.

E-books are already available on the internet and can be found with a simple search. Users value simplicity, and libraries are willing to offer quality e-books and to follow the law relating to e-books. It is in the hands of the different players involved to find and consolidate a new way of reading, teaching and learning. In the context of the Spanish implementation of academic e-books, there are important challenges, and business opportunities too (Cordón García, Alonso Arévalo and Martín Rodero, 2010).

UOC Websites

UOC, www.uoc.edu

UOC's educational model,
www.uoc.edu/portal/english/la_universitat/model_educatiu/introduccio/
index.html

UOC Library, http://biblioteca.uoc.edu/eng/index.html

UOC Library e-books site, http://biblioteca.uoc.edu/ebooks/eng/

UOC Office of Learning Technologies, http://lt.uoc.edu

References

Cervera, A. (2010a) Document Management in the Open University of Catalunya (UOC) Classrooms, *D-Lib Magazine*, 16 (7–8).

Cervera, E. (2010b) The Deployment of an E-book Consultation and Loan Service: the experience of the Open University of Catalonia Library. In *Jornades Catalanes d'Informació i Documentació (12es: Barcelona: 2010), 19–20 May*, Collegi Oficial de Bibliotecaris Documentalistes de Catalunya (COBDC).

Cordón García, J. A., Alonso Arévalo, J. and Martín Rodero, H. (2010) The Emergence of Electronic Books Publishing in Spain, *Library Hi Tech*, 28 (3), 454–69.

Gros Salvat, B. (2009) *The UOC's Educational Model: evolution and future perspectives*, UOC.

Johnson, L., Smith, R., Willis, H., Levine, A. and Haywood, K. (2011) *The 2011 Horizon Report*, The New Media Consortium, http://net.educause.edu/ir/library/pdf/HR2011.pdf.

March Mir, F., Vállez Letrado, M., Benítez Juan, B. and Leg Gil, M. (2010) Revista de Biblioteconomia i Documentació. In *L'Evolució de la Biblioteca de la UOC: perspectives de futur*, 52, June 2010, 37–49.

6

Mobile service providers and library services in a multi-campus library

Ela Volatabu Qica

Introduction

Mobile technology has taken information and communication technology (ICT) around the world, the Pacific included, to greater heights. People with mobile technologies are able quickly to communicate, gather and share information without the limitations of time and place that once existed. The technology has revolutionized and even overwhelmed people's lives, altered the way normal business is done and may have increased expectations or fears of what lies in the future. While we are mapping out the best strategies for our 'niches', the mobile service providers are busy marketing their chosen products. The onus is on librarians, as information specialists, to embrace the new mobile technologies in their work and to make their demands known with regard to the use of mobile technology so as to address geographical areas where services are lacking.

This chapter attempts to address the mobile broadband and text service options offered by the two major service providers, Vodafone and Digicel, in Fiji, and highlights implications for the role of librarians in the Pacific. It begins with an overview of the region, the University, information technology and the library so that the diverse physical, economic, social and cultural dimensions that we must work with can be appreciated.

The Pacific region comprises 30 million square kilometres of ocean, making it three times greater than the area of Europe. The total land mass, however, is almost equal to that of Denmark. The region's territorial divisions are Melanesia, Polynesia and Micronesia and it is home to one-third of the world's languages. The total number of inhabitants is about 1.3 million, which is 0.1% of the world's population. Despite the varying degrees of diversity within the region, the introduction of mobile technologies has brought about changes in communication and also produced a current generation of information technology (IT) savvy clients, also referred to as 'Net Gen' users – a term coined by Oblinger and Oblinger (Lippincott, 2008) – a majority of whom are patrons of the University of South Pacific (USP) Library.

University of the South Pacific

Established in 1968, USP is one of only two universities of its kind in the world and is jointly owned by the governments of 12 Pacific island countries (Cook Islands, Fiji, Kiribati, Marshall Islands, Nauru, Niue, Samoa, Solomon Islands, Tokelau, Tonga, Tuvalu and Vanuatu). The University is governed by its own Council and a Senate that is responsible for matters such as teaching and research. It offers programmes through distance and flexible learning in a variety of modes and technologies throughout its 14 campuses. Both international and regional students and staff give the USP a multicultural character. The University's vision is 'to be a university of excellence, highly regarded locally, regionally and internationally' (University of the South Pacific, 2009). Underpinning this notion of excellence are the USP's pillars of Quality, Relevance and Sustainability.

The USP network, USPNet, is a wide-area network incorporating 5MHz IP satellite-based technology that connects all regional campuses in member countries with the main campus in Laucala, Fiji. This technology provides for interactive audio tutorials, e-mail communication, access to the world wide web, live video and multi-media material. However, the technology has the limitation that it connects only to campuses and not directly with individual students remote from the campus vicinity. It is through this network that the USP Library carries out satellite meetings with its counterparts on other campuses each month.

The USP Library

The main USP Library is located in Suva, Fiji, on Laucala Campus and is fully automated. Through the USPNet, both local and regional students are able to access the online public access catalogues (OPACs) of three major campuses (Vanuatu, Samoa, Fiji – USP also has libraries in other parts of Fiji: two campus libraries, Lautoka and Labasa, and two sub-campuses, Sigatoka and Rakiraki). Currently all library services to students, staff and researchers operate through a combination of manual and automated processes. While items are issued online, books are still stamped with due dates and users need to visit the premises physically to borrow and return books. Reservations are placed via the OPAC and a courtesy service is provided to staff to receive renewal, overdue and recall notices.

As the trends for change occur in other libraries and case studies amass literature on mobile libraries, it is imperative for the USP Library to make use of appropriate available mobile technologies to strengthen its role in meeting the needs of current and future generations of learners. In order to achieve this, we need to consider the work of mobile service providers in Fiji.

Mobile service providers

The major mobile service providers in Fiji are Vodafone and Digicel. Exclusivity licences existed until regulatory changes were made through the government's Telecommunications Bill in 2007. Prior to this, Vodafone Fiji was the sole mobile service provider, but lost its monopoly status when Digicel Fiji established its network in 2008. In a survey conducted by Fong et al. (2010), it was found that 78.9% of students subscribed to Vodafone, making the company still very dominant in the industry. This percentage may change in the future if new regulations are considered. However, 'further reform to telecommunications regulation is unlikely while Fiji is under interim military rule' (BusinessWire, 2010).

We will now consider the options that can be provided by the mobile service providers for the USP Library. Two of the many services have been identified: mobile broadband and mobile phone. What is mobile broadband? It is also known as wireless broadband and makes possible wireless transmission at high speed via a connection card inserted into a computer or mobile router. In Fiji a common format used is a USB or a 'dongle'. As a product, it is 'Digimodem' to Digicel and 'Flashnet' to Vodafone. These gadgets allow users access to the internet and e-mail while on the move. In their terms, it is described as 'the internet in your pocket' (Vodafone Fiji, 2011), making the technology both flexible and mobile. Currently, USP students with laptops who can afford the Flashnet are already using the gadget because internet access via the PCs that are available to students in the Library or the computer labs is always fully in use. Through this gadget, students can access the Library OPACs from the web and view their borrower details, number of books loaned, books overdue, fines and reserved books available for collection from the Library. The maximum tariff plans of both Digicel and Vodafone are shown in Table 6.1. As prices are maximum-level plans, we cannot determine what costs students are paying for the service unless a survey is carried out.

Table 6.1 Mobile broadband tariff plans

Company	Plan	Data cap	Subscription monthly	Cost excess data
Vodafone	Broadband Mega	20 GB	$80.95	$0.10 per 10 MB
Digicel	Data Premium	8 GB	$65.00	$0.03 per 1 MB

While broadband has opened up access to the internet and e-mail, the mobile phone has taken information sharing to another promising level. Fong et al.'s 2010 survey showed that 93% of students had phones. The majority of these students will be those 'who have grown up with portable audio devices . . . and accustomed to "grab and go"' (Lippincott, 2008). The mobile phone can link librarians directly with individual students.

The SMS bulk messaging via mobile phone is one feature to be considered. Digicel and Vodafone both provide this service, whereby one message limited to 160 characters can be sent to a block of users. Messages ranging from library alerts to opening hours and library emergency announcements can be sent to each student in this way, but a deciding factor for whether this service is likely to be implemented is affordability. How much would it cost the USP Library in Fiji? Table 6.2 provides details of a price plan from Digicel.

According to the USP Annual Report 2009, the estimated number of full-time students in Fiji stood at 12,451 and the grand total for the

Table 6.2 Price plan from Digicel for USP Library's SMS bulk text messaging service

Msg per mnth	Wholesale price (FJD$/SMS)		
	Digicel	Other networks	International
0–5000	0.15	0.15	0.15
5,000–15,000	0.13	0.15	0.15
15,001–30,000	0.11	0.13	0.13
30,000–100,000	0.10	0.13	0.13
100,001–200,000	0.10	0.10	0.10
200,001+	0.10	0.10	0.10

region at 18,621. With reference to Table 6.2, for the range of 5000 to 15,000 messages, eight bulk messages per month to students in Fiji would cost FJD1.04 (8*0.13). As greater numbers of corporate organizations use this service, the quoted price will surely come down, allowing us to send messages to the regional campus students also. It is hoped that in the not too distant future the USP Library will be using one of these services for direct communication to students. If this is our dream for enabling library services through mobile technology to the students in Fiji and other regional campuses, what implications will it have for librarians in the USP region?

Implications

Librarians will need to enhance their skills and raise their awareness of the use of these mobile technologies so as to be conversant with the tools when students need assistance. This is possible through close networking with mobile service providers Vodafone and Digicel, attending and participating in forums such as the M-libraries Conference that address the subject, and using the technology so that one knows and is familiar with how it operates.

In terms of delivering content, librarians will need to know what particular library service users want to access and read via a mobile phone, and what types of user these are likely to be. Noting that the use of mobile phones is currently prohibited in the USP Library, it will be imperative for librarians to discuss this matter and perhaps provide a 'commons' or 'special' reading room.

Librarians should be prepared to work in both paper and digital environments. This means that a librarian will be expected to multi-task more and also to become 'mobile'. For example, although one can be a reference librarian to an on-campus student, one will also be expected to instruct a student off-campus in how to use the mobile phone or mobile broadband to access library services.

Librarians will need to be more assertive and bolder in the way they use mobile technology for their work. This means that they should be making their demands known to the mobile service providers in the areas that either lack service or are in need improvement.

Conclusion

For the USP region, the use of mobile technology in libraries is in its early stages. Most libraries lack the services of full-time professional staff and have other operational priorities on their agendas that need attention. There is not going to be a better time than now to gradually begin using mobile technology to offer library services to students. It is hoped that the digital divide will be narrowed once a choice and implementation for either mobile broadband or mobile phone is made – first for the campuses at Laucala, Fiji; Emalus, Vanuatu; and Alafua, Samoa; and then for the other campuses.

References

BusinessWire (2010) *Research and Markets: Fiji – telecoms, mobile and broadband*, (16 September), www.businesswire.com/news/home/20100916005792/en/Research-Markets-Fiji.

Fong, E. et al. (2010) *USP Mobile Learning Survey for Students*, unpublished.

Lippincott, J. K. (2008) Libraries and Net Gen Learners: current and future challenges in the mobile society. In Needham, G. and Ally, M. (eds), *M-Libraries: libraries on the move to provide virtual access*, Facet Publishing.

University of the South Pacific (2009) *Annual Report*, USP.

Vodafone Fiji (2011) *Marketing Website*, www.vodafone.com.fj.

7

Using mobile technology to deliver information in audio format: learning by listening

Margie Wallin, Kate Kelly and Annika McGinley

Introduction

Libraries globally are exploring ways to deliver the myriad of quality information resources available to their users – via mobile devices. As well as focusing on documents (with the inherent problems associated with reading on small screens), Southern Cross University (SCU) Library has explored ways of accessing key documents and converting them to audio form, so that students and staff have an alternative way of engaging with academic literature.

Learners today exist in an increasingly multimodal environment. Mobile devices provide them with extra levels of flexibility, offering them both auditory and visual input and allowing them to learn anytime, anywhere and while doing anything. As Eisenwine and Hadley (2011, 5) state, 'the digital generation prefers parallel processing and multitasking as a way of digesting information. In addition, that generation prefers pictures, sounds and video over text.'

Traditionally, academic libraries have provided access to predominantly text-based materials. With a highly mobile cohort, however, the problems inherent in viewing or reading on small screens, combined with the need to support different learning styles, have provided the impetus for exploring audio-based alternatives to the academic literature.

As SCU Library staff worked towards the development of a mobile platform, they simultaneously explored the provision of text-based audio files within that platform and the University's learning management system.

Read, listen and learn

The pedagogical value of providing audio-based resources, especially podcasts, has been well documented (Hew, 2009; Kervin et al., 2006; McGarr, 2009; Tempelhof, et al., 2009). Generally, such podcasts have been based on lecture materials or supplementary materials designed to support specific units. Other examples include the provision of conference or workshop presentations. Little has been written evaluating the use of audio forms of published and/or scholarly literature.

Between 2008 and 2010, database vendors began providing audio files of selected academic literature. Factiva led the field with linked audio files for all its articles (of less than 4000 words). EbscoHost and Gale Cengage followed suit, providing audio files for full-text HTML documents, but not for PDF files. All three vendors provided an immediate 'listen' function and the facility to download an MP3 version. Ironically, despite being the provider of the first mobile database platform, EbscoHost disabled the 'audio' function for mobile devices. Ebook Library (EBL) provided a 'read aloud' facility within its e-book collection, but not downloadable MP3 files.

In January 2011, SCU Library invited the University community to participate in a research project to evaluate the use and usability of audio files as learning tools. The 'Mobile Resources' Library guide (http://libguides.scu.edu.au/mobile) incorporated a research project tab that linked to selected journal articles (on m-learning/e-learning) with both text and audio formats. A screencast provided instructions on downloading and listening to the selected articles and a link to the survey tool was provided.

Forty-two responses were received, from both staff (15%) and students (85%). This low response rate was not unexpected because the project was undertaken during the traditionally quiet summer session of the academic year. All faculties and schools within the University were represented, and there was a fairly even spread across ages, which ranged from 18 to 64 years. Of the student cohort, 41% were studying internally, 6% were offshore students and 53% were studying at a distance.

Survey participants were asked to comment on how effective they found a particular format for their learning, and which article format they would prefer to access on their mobile devices. A preference for access to both formats was expressed by 64% of respondents. Many of the comments reflected the general advantages of mobile devices – convenience and flexibility:

> It's a lot easier than lugging around a huge text book or pages that add to clutter and can get lost.
>
> More mobile than sitting at a computer work station. Was able to relax in own environment.

Not surprisingly, the main criticisms of accessing 'text' on a mobile device related to screen size and the associated reading difficulties. Some respondents also commented on the limited or non-existent 'note taking' functionality of some devices:

> Words too small to read.
>
> I like to use the hard copy so I can make notes and add Post-its.

Having audio access to journal articles was a new feature for many of our survey respondents, and this was reflected in their overwhelmingly positive comments:

> As I am an international student, this format assists me to develop my listening skill.
>
> It was excellent. How good, can we have all the journals available to listen to?
>
> I prefer reading for academic-related info, however I found podcasts useful. I would download audio to my mobile device and play it while driving, or maybe peeling carrots, etc. . . . This sort of facility would be very useful for me.

However, despite the positive responses, almost all criticisms of the audio format pertained to the voice quality – its synthesized nature and speed of delivery:

> But I want a human voice rather than a computer-generated one.
>
> I found I was able to read much quicker than the speaker was relaying the text, however, with the amount of reading time available to us, as it takes total concentration, being able to listen while walking, driving, on the train etc. has huge benefits. Having the text means you are able to refer back to specifics if needed.

Having access to both text and audio formats was the most popular approach (64%):

> This is such a great idea and I hope the time and effort is put in to make this a freely available tool for all students. Accessing the data in both audio and text format is so invaluable as a greater understanding is gained by using both styles of learning. It is a far more complete experience.
> Although I did notice that when I read the article after listening to it I picked up some things I hadn't when just listening.
> Listening is good but reading can allow clarification.

A preference to access neither format on a mobile device was expressed by 10% of respondents. Generally these comments reflected a preference for print materials:

> With a difficult article I need to study what is being said with frequent backtracking. I usually prefer to print the difficult papers.

Unit-based cohort

In addition to the resources made available to the entire University community, selected articles – relevant to two assessment tasks – were embedded in online course material for a postgraduate unit (Foundations of Management) within the School of Business. This cohort was keen to embrace new learning technologies – being time poor, yet with ready access to technology, internet and mobile networks. Students provided informal feedback via a virtual focus group within the learning management system.

Feedback both on the forum and via the discussion lists indicated a lack of expertise in downloading (usually via iTunes) and managing the files. Students discussed technical frustrations that diminished their experience. In response, the SCU Library and Flexible Learning Development Services (FLDS) are developing a 'software toolkit' with advice and screencast instructions on managing multimedia files.

Students on the forum were enthusiastic about resources provided in multiple formats for use on mobile devices. One student explained that he was able to expand his study time by listening to audio content on his daily commute to work. For these students, the overwhelming response was that providing material in multiple formats allowed them to make better use of their time and to use different formats to optimize 'reading' time.

Audio investigations

As a result of this feedback, SCU Library staff continued to investigate opportunities for providing both text and audio forms of academic literature to their users. Feedback was provided to those database vendors who offer an audio alternative to the academic literature, based on respondents' comments. The provision of such services by other database vendors is being monitored on an ongoing basis and new resources are promoted when available. Journals providing their own podcasts (e.g. *Nature, Wall Street Journal* and *New England Journal of Medicine*) have also been explored and promoted (Tempelhof et al., 2009).

In addition to providing access to existing audio files, the Library hoped to generate audio files (using text-to-speech software) for unit e-reserve materials. However, there are a number of implications pertaining to the creation and distribution of 'adaptations' of such literature. Each publisher or vendor applies varying restrictions or negotiates specific licence agreements, and having to investigate this on a document-by-document basis would not be cost or time efficient. In one study (Miller and Piller, 2005) audio versions of set readings were created. There were a limited number of readings involved and permission was sought for each reading.

While institutions are bound by copyright and licensing restrictions, individuals are able to create such files, solely for personal research or study. It was decided to explore the possibilities of making text-to-speech tools available to university students and staff.

Text-to-speech

Text-to-speech tools (or speech synthesizers) have been in use for many years, often for accessibility purposes or to support language students. Such tools are equally valuable for auditory learners, and some interesting applications of these tools include essay revision, assisting struggling readers and providing narration for web lectures (Balajthy, 2005; Chong, Tosukhowong and Sakauchi, 2002; Garrison, 2009).

As stated by Rughooputh and Santally (2009, 137), 'Research into personalization and individual learning preferences has shown that the use of multi-modal approaches to delivery can help improve learning experience of learners irrespective of whether in classrooms or through distance education.' This reinforces the comments made by a number of the survey participants.

A range of free and commercial text-to-speech software tools and applications were identified and evaluated. Evaluation criteria included ease of use, voice quality, choice of accents and speed, file size and format, as well as cost. As with the audio files provided by database vendors, the quality of 'voice' was viewed as a key consideration. A selection of these tools were documented, along with key features and comments, and the information was made available to SCU staff and students on the Text-to-Speech tab within the 'Mobile Resources' Library guide (http://libguides.scu.edu.au/mobile).

As mobile devices are more than just 'phones', desktop-based software and mobile applications were also investigated. Thus, all users are supported – whether they have high-end mobile devices, including smartphones, or just basic MP3 players. This ensures a service that is both equitable and flexible. One comment from a student was: 'I will still access files via the web and then load them onto my phone from my computer (it is cheaper that way)'.

Initially, information about text-to-speech resources was provided to participants in the research project and promoted via the SCU Library web page. Individual feedback was extremely positive. Students and staff reported using these tools to create MP3 files of journal articles, selected e-readings and course materials, book chapters and book sections, web pages, newspaper articles – to name just a few. All the respondents commented on the value of these tools and the increased flexibility they provided to them as mobile learners. They also reinforced the comments from the initial survey, on the value of having multiple formats of the same resource.

Conclusion

Novelty is probably one of the most powerful signals to determine what we pay attention to in the world. (Poldrack, 2010, 1)

While the use of audio files or podcasts in the academic environment is not new, the ability to access audio forms of scholarly literature is a recent development. In addition, enabling staff and students to create their own personalized collections of audio files – generated from text-based resources – is both novel and empowering for mobile learners.

This project has demonstrated that scholarly literature can be repurposed to suit a variety of learning styles, particularly in the mobile setting. The success of embedding dual-format, mobile-ready scholarly resources into a unit has also created opportunities within other unit offerings.

In response to the research findings, the SCU Library, FLDS and the Teaching and Learning Centre continue to collaborate on the development of 'mobile literacy' resources, enabling students to navigate the technical complexities of creating or downloading and listening to content via mobile devices.

As Low and O'Connell (2006, 2) note, 'the highly personalised nature of digital mobile devices provides an excellent platform for the development of personalised, learner-centric educational experiences'. By providing opportunities for learners to both create and access multi-format and mobile-ready scholarly resources, the Library continues to support mobile (and indeed all) learners in their academic endeavours.

Acknowledgements

Trevor Davey, Lecturer, School of Business, Southern Cross University.

References

Balajthy, E. (2005) Text-to-speech Software for Helping Struggling Readers, *Reading Online*, 8 (4), 1–9.

Chong, N. S. T., Tosukhowong, P. and Sakauchi, M. (2002) Whiteboard VCR: a web lecture production tool for combining human narration and text-to-speech synthesis, *Educational Technology & Society*, 5 (4).

Eisenwine, M. J. and Hadley, N. J. (2011) Multitasking Teachers: mistake or missing link? *Educational Forum*, 75 (1), 4–16.

Garrison, K. (2009) An Empirical Analysis of Using Text-to-speech Software to Revise First-year College Students' Essays, *Computers & Composition*, 26 (4), 288–301.

Hew, K. F. (2009) Use of Audio Podcast in K-12 and Higher Education: a review of research topics and methodologies, *Educational Technology Research and Development*, 57 (3), 333–57.

Kervin, L., Reid, D., Vardy, J. and Hindle, C. (2006) A Partnership for iPod Pedagogy. In Markauskaite, L., Goodyear, P. and Reimann, P. (eds), *Proceedings of the 23rd Annual ASCILITE Conference, Who's learning? Whose technology?* Sydney University Press.

Low, L. and O'Connell, M. (2006) Learner-centric Design of Digital Mobile Learning. In *Learning on the Move: online learning and teaching conference*, Queensland University of Technology.

McGarr, O. (2009) A Review of Podcasting in Higher Education: its influence on the traditional lecture, *Australasian Journal of Educational Technology*, 25 (3), 309–21.

Miller, M. S. and Piller, M. J. (2005) Principal Factors of an Audio Reading Delivery Mechanism: evaluating educational use of the iPod. In Kommers, P. and Richards, G. (eds), *Proceedings of World Conference on Educational Multimedia, Hypermedia and Telecommunications*, AACE.

Poldrack, R. (2010) Novelty and Testing: when the brain learns and why it forgets, *Nieman Reports*, 64 (2), 9.

Rughooputh, S. D. D. V. and Santally, M. I. (2009) Integrating Text-to-speech Software into Pedagogically Sound Teaching and Learning Scenarios. *Educational Technology Research and Development*, 57 (1), 131–45.

Tempelhof, M. W., Garman, K. S., Langman, M. K. and Adams, M. B. (2009) Leveraging Time and Learning Style, iPod vs. Real-time Attendance at a Series of Medicine Residents Conferences: a randomised controlled trial, *Informatics in Primary Care*, 17 (2), 87–94.

8

Sound selection: podcasts prove positive

Daniel McDonald and Roger Hawcroft

Introduction

The Toowoomba Clinical Library has introduced and established a service whereby clinically oriented audio presentations are provided to the health professionals throughout the region it serves. This chapter describes its inception and progression.

In many ways, this is a project that is all about mobile information technologies – about MP3 files and podcasting software and iPod shuffles. In some crucial ways, though, the technology is almost the side issue, or at the very least was not the driver of the success that this service seems to have achieved. In his recent manifesto 'You are not a gadget' Jaron Lanier (2010) claimed 'The central mistake of recent digital culture is to chop up a network of individuals so finely that you end up with a mush. You then start to care about the abstraction of the network more than the real people who are networked, even though the network by itself is meaningless. Only the people were ever meaningful.' It is an obvious statement, yet a profound observation, and one of which the Library has tried to remain cognizant in all its work.

The major role of a clinical library service is to be aware of quality content and to connect clients – doctors and nurses and allied health workers – with it. It is they who make decisions about patient care, who strive to practise from a sound evidence base, who must provide safe, effective and cost-efficient health interventions. To do this, clinicians require information. The eminent UK physician Sir Muir Gray has consistently argued in many forums that in the 21st century knowledge is the key to improving health. In the same way that people need clean, clear water, they have a right to clean, clear knowledge. He has also claimed that the greatest future advances in healthcare will come not from new inventions and discoveries but from the application of what we already know. To that end, it is the work of the library and its information professionals to meet the need – to put the right information in the right context and the right container for the right client. This project is one small contribution towards meeting that need.

Where did we start?

For several years the Library had loaned first audio cassette and later audio CD programmes as a regular part of the circulation service. These programmes had long been used by clinicians when driving between regional centres or from rural locations to the city, as a convenient way of using their time productively and keeping up to date with developments in their field. With the increasing ubiquity of the MP3 player, either as a stand-alone or as part of smartphones and computers, for music and entertainment, we had begun to investigate whether there was a role for that technology in the information arena. A paper published in November 2007 by Ashok and Priya Roy seemed to reinforce our still nebulous thoughts. They quoted a Pew Internet study that found more that 22 million American adults owned iPods or MP3 players and that 29% of them had downloaded podcasts from the web. They explored a number of intersections of training and podcasting in adult education, including a programme at Duke University where all incoming first-year students were supplied with iPods, but much of their discussion surrounded possibilities rather than existing services.

It wasn't until a request from a palliative care doctor, one of Lanier's 'people who are meaningful', that our thoughts took a concrete direction. This doctor was travelling once a week between Toowoomba and Brisbane and wanted to redeem the time by listening to material relevant to his field. There was next to nothing available in established collections and, not wanting to return to the client empty-handed, Library staff began browsing the web, trying to find out what was available for health professionals. Amongst some obscure material were enough quality resources to convince this doctor of its worth, so once a week two or three audio CDs were created from MP3 files that had been found on the web and downloaded. These were inauspicious but important beginnings. Informal feedback from the palliative care doctor gives several clues as to why the project has become what it is now. He said the CDs were easy to listen to while he was driving on the highway. Some of the presentations were particularly instructive and he would listen to them several times, and would also ask for copies to be shared with colleagues. He did complain about the prevalence of American accents, so a lecture on telomeres was obtained from the Australian researcher Elizabeth Blackburn. Later on she was awarded a Nobel Prize for this work, so to have been exposed to her work beforehand was rewarding. From the Library's perspective, the experience proved to be mutually beneficial. The cost of sourcing the material was low and, through evolving search strategies, an increasing wealth of sources of freely available MP3 files relevant to our professional community was discovered.

What did we do?

In reflecting upon this work in team meetings we realized that this individual and very particular service was eminently scalable. It seemed clear that there was enough quality and relevant content available to be worth our investment of time and resources. It was important for our staff to know what like-practising clinicians throughout the world were saying. By the late 2000s shared and shareable information was no longer solely the domain of the printed word and the spoken word had been recorded and published online for some time. We also speculated that many busy practising clinicians would value listening to scholarly communication either instead of or alongside the written word: such information presented as conversation or lecture can be easier to digest and taps into modes of learning that differ from reading the printed or screened word.

We also suspected that most of the content was 'grey' and was not being listened to by those who could benefit from it the most. It is all very well for the cloud to be rich in atomized information, but unless the cloud condenses into rain and that information is accessible to those who want and need it, an unproductive drought and confusing fog will be all that ensue. We did not want to simply post a list of descriptive links on an intranet page somewhere, in the hope that some industrious soul would discover them and know what to do. All of our health professionals are good at what they are trained to do, but not all are tech savvy. Even those who are may not have the time and intellectual energy to access, download and play these presentations. It was our job to give them the information so that they could listen, learn and apply.

To this end we set about determining the best way to link our clinicians with this audio information. We realized that CDs had limitations and so we pursued the idea of purchasing iPods for loan. We understood that they might be an easy target for theft, but no more so than any mid-range text currently in our collection. Likewise, we hoped that acquiring bright and funky technology such as Apple's *sui generis* iPods and putting them into the pockets and ears of borrowers would help to counter the misconception that libraries were simply mausoleums for unread books. So we bought ten iPods and spent some time doing nothing with them. Again, it was Lanier's insight into 'people who are meaningful' that gave our ideas and the technology the direction and focus it needed.

We focused our attention on junior doctors. They rotate through the major clinical areas of the hospital, so building a collection of podcasts around their needs would ensure that most of the clinical disciplines were addressed. We imagined residents or interns starting a new rotation by coming to borrow an iPod dedicated to orthopaedics, then emergency medicine on their next rotation, and so on. The iPods would be pre-loaded with audio material relevant to each particular discipline. They would simply borrow, listen, learn, return, repeat. We knew that not all the material would be directly applicable to their circumstances and abilities, but enough would, and sometimes it is good to be stretched, and exposed to exotic instruction – and one press of the skip button was always available. We also supposed that junior doctors would be the most amenable to adopting 'new' technology, and since many consider the state capital, Brisbane, home, they often have travelling time to redeem and study burdens that are great. In adopting this approach we also knew that having subject-specific iPods would mean that consultants or nursing staff who specialized in those areas would find the players just as useful.

We had an audience. We had players – wee things in lime green and brilliant fuchsia and electric blue tucked into white boxes with white earbuds (we clean them on each return) – accessioned and added to the catalogue and available for circulation. What we needed was content. This we built through intensive searching and much browsing, trying a variety of techniques and terms, and learning where sites typically post their audio content. In the course of this searching we have assembled a sizeable list of useful sites and are endeavouring to make this targeted collection available to our clients and others through a Delicious page or similar. The vast majority of material collected has been freely available on the web, but we do maintain a small number of subscriptions to commercial audio medical education providers who regularly deliver material in MP3 format. Typically, we download MP3 files rather than subscribe to podcasts, as this gives us greater flexibility and allows us to filter unwanted material. We are conscious of copyright issues and have sought permission where possible and appropriate, and have been uniformly delighted with the support received. The collection is not static, and we are continually adding to the files as new material is posted or as we come across new troves. The type of material we download varies greatly and includes specialist presentations, summaries from key journals, lectures from conferences and specialist societies, continuing medical education sessions and interviews with leading researchers. By and large, the content is rich in diversity, of good quality and well presented.

What have we achieved?

Our hunches and our investment paid off. In mid-2008 we announced the availability of the iPods through a viral marketing campaign that generated substantial interest. We were worried that the idea would flare and then die away quickly, but this has not proved to be the case. At least 50 staff have borrowed one of the ten iPods that the Library has populated with subject-specific content and made available for loan. One fact that we were initially slow to grasp but have quickly capitalized on is the number of staff who have their own MP3 players, particularly with the proliferation of smartphones. Consequently, it is the content alone that they are interested in. With this access mode in mind we built a database that not only helps to track the collection but also allows staff to generate their own 'playlist'. It is similar to iTunes but much more customizable to our needs. This has proved immensely valuable. Over 100 additional clients have received content independently of our iPods, either via a library-generated CD (for which there is still demand) or by transfer to their own player. Effectively, several thousand audio presentations have been listened to. Feedback received informally and through a structured survey has been uniformly positive, with comments typically highlighting the ease of access and the quality and relevance of the material provided. Anecdotal evidence suggests that several presentations receive multiple plays or are subsequently given to colleagues. Further, increasingly, interactions with this content are not static but iterative. After initial exposure clients are coming back and asking for more material in their field, or more material in a different field, or much more heavily filtered material, or material useful for continuing professional development schemes. In addition, specialist collections have been placed in specific department areas to allow staff easy browsing and access. This has proved immensely popular, especially the emergency medicine podcasts in rural facilities and World Health Organization infectious disease webinars, and our local infectious disease unit and recorded telephone education workshops in cancer care. This extends the scope for librarian–clinician collaboration, which can only improve the nature of the service and enhance the involvement of the library in the clinical community.

Opportunities for inter-organizational collaboration have also arisen through engagement with Greenslopes Private Hospital Library (Brisbane) and Queen Elizabeth Hospital Library (Adelaide). These opportunities came about through publication of the project in *Incite* and *HLA News*. The audio collection and model of service have been shared, while methods for comparative evaluation are being explored. To have involved other health libraries with many similarities and some key differences not only increases numbers for and awareness of this project, but opens possibilities for other areas of collaboration. This is tremendously exciting. It is also encouraging to note that this project was runner-up in the 2010 ALIA/IOG Excellence award and the recipient of the 2011 HLA/HCN Innovation prize. Such peer recognition is flattering, and also is well received among the decision makers and budget controllers of the hospital executive, giving them confidence to support future innovation by the Library.

Conclusion

To reiterate, this project was never solely about iPods or podcasts. As a library, we have a responsibility to ensure that the clinicians whom we link to information have access to the best available evidence and swiftest translation of bench-to-bedside research. This project is our attempt to honour one part of that responsibility. Through this project, clinicians in Toowoomba have had the opportunity to listen to lectures they would never otherwise have heard. This has expanded the knowledge base of the local clinical community, which in turn results in better-informed decisions for patient care. Ultimately, patient care is where health-library best practice should be measured, and it is where we believe that this project has made a significant contribution.

References

Lanier, J. (2010) *You Are not A Gadget: a manifesto*, Alfred A. Knopf.

Roy, A. K. and Roy, P. A. (2007) Intersection of Training and Podcasting in Adult Education, *Australian Journal of Adult Learning*, 47 (3), 479–91.

Part 2

People and skills

9

Staff preparedness to implement mobile technologies in libraries

Sarah-Jane Saravani and Gaby Haddow

Introduction

Mobile technologies have added a new dimension to the role of library staff, a role that requires relevant knowledge and skills to ensure that the needs and expectations of clients are met (Gentry, 2011; Kroski, 2008; Traxler, 2008). For libraries to succeed in the mobile environment, management will have to consider two significant staffing issues. Firstly, it needs to understand the knowledge and skills required by staff to enable them to deliver services through mobile technologies; secondly, that understanding should be applied to develop support and training for staff in the use of mobile technologies. In order to explore these issues a survey was undertaken in late 2010, involving librarians working in the vocational education and training (VET – similar to further education in the UK) sector in Australia and New Zealand.

Vocational education and training libraries and mobile technologies

Generally, research into mobile technologies across the library sector in tertiary education has focused on universities (Adams, 2009; Kealy, 2009; Zauha and Potter, 2009). While there are similarities between university libraries and VET libraries, the differences are notable. There are 58 VET institutions in Australia and 19 institutes of technology and polytechnics (ITPs) in New Zealand. In 2008 enrolments numbered 1.7 million for Australia and 78,000 in New Zealand (Commonwealth of

Australia, 2009; New Zealand Ministry of Education, 2009). These are large and diverse groups, comprising students taking apprenticeships, certificate, diploma and degree-level courses. The sector is characterized by a focus on industry and community engagement, whereby students tend to be competency based rather than research oriented and many have ongoing workplace commitments related to their studies. Prolonged periods of absence from campus and the young average age of the student cohort mean that effective mobile library services are particularly critical for the VET sector (Douch et al., 2010).

About the study

Library staff from eight New Zealand and six Australian VET institutions were invited to participate in a survey that included a short online questionnaire and an interview (conducted in person, over the telephone or via Skype). The questionnaire gathered some demographic information and asked participants to indicate their competence in using mobile technologies. The interviews used open questions to discuss the knowledge and skills required to deliver mobile services and the on-the-job training required to acquire those knowledge and skills. Purposive sampling enabled the recruitment of three professional staff from each institution, representing a management position, an ICT-related position and a position without either of the former responsibilities. Forty-two librarians participated in the survey, with a slight skew in the sample towards management positions (16 participants) and equal numbers of ICT-related and general librarian positions (13 participants for each position type).

Competence in using mobile technologies

In order to establish the competency of library staff in relation to mobile technologies, participants were asked to indicate their level of competence on a five-point scale, from beginner to advanced. This scale was then collapsed into a dichotomous measure of competent or not competent. The findings from the 40 participants who responded (Table 9.1) show that the staff in management and ICT-related positions reported higher levels of competence. Overall, more than half of the sample reported competence in using mobile technologies.

Table 9.1 Competence in using mobile technologies by position type (n=40)

Position	Competent		Not competent	
	No.	%	No.	%
Library manager	9	22	5	13
ICT librarian	8	20	3	8
Librarian	6	15	9	22
Total	23	57	17	43

The self-reported competence levels were examined against the number of years participants had worked in libraries. Although the results suggested an association of shorter length of service with competence, the number of participants in each service length range was too low to reach a firm conclusion. Table 9.2 presents the findings as raw numbers only.

Table 9.2 Competence in using mobile technologies by length of service

Length of service (yrs)	Competent	Not competent
1–5	3	1
6–10	3	0
11–15	6	0
16–20	4	6
21+	7	9
Total	23	16

Knowledge and skills required by staff delivering mobile library services

The qualitative data gathered in the interviews, using a recording device and notes, was analysed through an iterative process that involved listening to the recordings and reading the notes repeatedly to identify categories and overarching themes. Participants' responses to the question relating to the knowledge and skills required for mobile library services delivery produced over 40 different types of knowledge and skills. The five most frequently cited of these categories are:

1 Staff access to and competence using a range of devices (n=19), expressed by participants as: 'We need to have our own mobile unit – tablet, smartphone – to use to see how the services work' and 'It would work the best if the staff were up-skilled as users of the devices'.

2 Willingness to try new technologies (n=11), illustrated by the
   following comment: 'a willingness to press the buttons . . . a
   willingness to jump in and get your feet wet'.
3 Knowledge of students' use and expectations of mobile technologies
   (n=10), as in: 'staff would need to be knowledgeable in what
   devices students are using'.
4 Skills to enable delivery of services through mobile technologies
   (n=9), as seen in the response: 'the library will need to adopt the
   technologies we want to promote our services on and become
   champions of this if we are to promote mobile technologies'.
5 Knowledge and ability to recognize opportunities for using mobile
   technologies (n=8), with comments such as: 'able to identify a
   need and then find a technology to meet that need and then
   learning from that'.

Categories less frequently discussed included knowledge of operating
systems (n=5), web-based technology skills (n=2) and an
understanding of compatibility issues (n=1).

In order to understand how these categories of knowledge and skills
might inform managers in the planning of mobile library services,
further analysis was carried out to establish overarching themes. Three
themes emerged:

■ technical – defined as information technology, software or hardware
  skills
■ management – defined as facilitating the attainment of
  organizational goals
■ adaptability – defined as the ability to engage with work-related
  opportunity.

These themes were cross-tabulated with position types to determine
whether a participant's role was associated with their perceived
knowledge and skills needs. The findings are presented in Table 9.3 and
indicate that most participants, regardless of position, see management
as central to attaining the knowledge and skills needed for the mobile
environment. It is evident from participants' responses – for example,
'there are other developers within the institution ... we are not on our
own in terms of identifying some of the needs' – that this responsibility

Table 9.3 Knowledge and skills themes by position type (n=42)			
Position	Technical	Management	Adaptability
Library manager	11	13	9
ICT librarian	7	12	9
Librarian	7	11	6
Total	25	36	24

is discussed in broad terms rather than focused on library managers only. Also notable is that the technical theme was reported by the fewest ICT librarians and only seven of the general librarians. In contrast, more library managers perceived technical aspects as important to knowledge and skills.

On-the-job training required

When participants were asked to describe the types of on-the-job training that were required in order to acquire the knowledge and skills to effectively develop and deliver mobile services, 39 training categories were identified in the responses. The five most frequently discussed were:

1 Hands-on experience with range of mobile devices (n=13), illustrated by comments such as: 'informal, "get your boots on" type of training' and 'we need time to play with the device and see how the functions work'.
2 Training in e-book readers (n=9). A specific training need in the use of e-book readers was evident in responses, for example: 'training on how an e-book works and how to download the programs for it'.
3 Gaining knowledge of applications for mobile devices (n=8). Participants were interested in library-related applications that could augment existing services, illustrated in the comment: 'iPod/iPhone – what these can do, what they mean. Including a look at some popular and library applicable applications'.
4 Creating mobile-friendly web pages (n=8). Some participants mentioned their IT department in relation to implementing mobile-friendly web pages, with comments such as: 'it is a matter

at the moment to convince the IT department to get the web page mobile friendly, this is a priority'.

5 Lack of current training and/or plans to introduce mobile services (n=6). Responses in this category varied from a lack of appropriate staff for training delivery to no plans for the introduction of mobile services, evident in: 'on the job training – at the moment, there is none . . . there are no formal training or education programmes for mobile' and 'I can't say what specific training we would need as we have no specific plans at the moment to implement any new mobile services'.

The 39 categories identified in the responses about on-the-job training were analysed further to draw out overarching themes. Three themes emerged:

■ technical – defined as information technology, software or hardware skills
■ service delivery – defined as enabling the provision of services to clients
■ competence – defined as demonstrated workplace-related ability and understanding.

These are presented in Table 9.4 and show that more participants noted training needs related to competence than related to service delivery. However, both exceeded the number of participants who mentioned technical training. There is remarkable agreement between the different position types in terms of the training needs for developing and delivering mobile services.

Table 9.4 Training themes by position type (n=42)

Position	Technical	Service delivery	Competence
Library manager	5	11	11
ICT librarian	6	8	10
Librarian	5	8	8
Total	16	27	29

Implications for training VET library staff

In relation to their self-assessed level of competence in using mobile technologies, library managers and ICT librarians are more likely to report existing competence than are their counterparts in general positions. It is unclear why this is the case, when staff in general positions work directly with students and might be expected to have more exposure to mobile devices than do their managers. A possible explanation is that general librarians have a better understanding of what they do not know about mobile technologies and therefore underestimate their competence. Alternatively, library managers may perceive competence in a broader context and report it based on their personal use of mobile devices. Almost half the sample believed that competence in using mobile devices was a necessary attribute for librarians delivering mobile services, and close to a quarter of the sample said that hands-on training in using mobile devices should be provided. This finding is evident in the most frequent response category for the two questions 'What knowledge and skills are needed?' and 'What on-the-job training is required in the mobile library environment?' It demonstrates the importance placed on having physical access to mobile technologies for developing competence and suggests that hands-on experience is imperative for libraries engaged in or considering the implementation of mobile library services.

Whereas professional development of this nature requires the support of library management, the second most common knowledge and skill category reported by participants – willingness to try new technologies – relates to the personal traits of library staff. This finding indicates a perception, among these librarians at least, that some staff are reluctant to engage with new technologies. However, if staff were to get hands-on experience with mobile devices this apparent problem might be resolved and having opportunities to learn about student use of mobile devices and the available applications (the third most frequent category of response to each question) could foster ability to recognize the potential for mobile technologies (the fifth most frequent category of response to the first question).

E-book readers are seen as important technologies in the library – recognition perhaps of their increasing use and of the complexities of providing access to resources through them. Hands-on experience and training in the use of e-book readers is a relatively inexpensive undertaking for library management and could be incorporated into training programmes.

The findings suggest that much of the responsibility for developing staff knowledge and skills in using mobile technologies rests with library management. While responses were evenly distributed between the themes of technical skills and adaptability, over three-quarters of the sample described the knowledge and skills required for the delivery of mobile library services in terms of facilitation by management. In the training themes that emerged there was less focus, however, on technical skills. Instead, the participants believed that demonstrated ability and understanding (competence) and enabling the provision of services (service delivery) were central aspects of their training for the mobile environment.

Conclusion

A number of the study's findings indicate ways in which library managers can enable the professional development of their staff in the mobile library environment, beginning with providing access to and training in a range of mobile devices. A focus on service delivery is implicit in several of the most frequent response categories and reflects the traditional role of library staff, regardless of the technologies or processes involved, which is to recognize their clients' needs and respond appropriately.

Finally, at least two of the institutions included in the survey had either not implemented mobile services or had no staff training in mobile technologies in place. Given the extent of mobile device use, failure to address the needs of clients in a mobile environment potentially makes these libraries ill-equipped and, for a generation of Google users, irrelevant to their clients.

References

Adams, C. (2009) Library Staff Development at the University of Auckland Library – Te Tumu Herenga: endeavouring to 'get what it takes' in an academic library, *Library Management*, 30 (8/9), 593–607.

Commonwealth of Australia (2009) *Australian Vocational Education and Training Statistics: students and courses 2008*, National Centre for Vocational Education Research, www.ncver.edu.au/statistics/vet/ann08/students_and_courses_2008.pdf.

Douch, R., Savill-Smith, C., Parker, G. and Attewell, J. (2010) *Work-based and Vocational Mobile Learning: making IT work*, Learnng and Skills Network, https://crm.lsnlearning.org.uk/user/login.aspx?code=100186&P= 100186PD&action=pdfdl&src=WEBGEN.

Gentry, M. (2011) Handheld Mobile Device Support and Training, *Information Outlook*, 15 (1), 16–19.

Kealy, K. (2009) Do Library Staff Have What It Takes To Be a Librarian of the Future?, *Library Management*, 30 (8/9), 572–82.

Kroski, E. (2008) On the Move with the Mobile Web: libraries and mobile technologies, *Library Technology Reports*, 44 (5), 1-48.

New Zealand Ministry of Education (2009) *2008 Tertiary Education Enrolments*, www.educationcounts.govt.nz/publications/tertiary_education/42244.

Traxler, J. M. (2008) Use of Mobile Technology for Mobile Learning and Mobile Libraries in a Mobile Society. In Needham, G. and Ally, M. (eds), *M-Libraries: libraries on the move to provide virtual access*, Facet Publishing.

Zauha, J. and Potter, G. (2009) Out West and Down Under: new geographies for staff development, *Library Management*, 30 (8/9), 549–60.

10

Apps and attitudes: towards an understanding of the m-librarian's professional make-up

Kate Davis and Helen Partridge

Introduction

This chapter will discuss a research project that identifies the skills, knowledge and attitudes of the m-librarian. Six library and information professionals engaged in the provision of m-library services throughout Australia were interviewed. Six themes emerged as being critical for the m-librarian: technology, personal traits, user focus, communication, collaboration, and research and development. The research is significant because it establishes an open dialogue between current industry professionals, library science educators and the professional association on the evolving skills and knowledge required by information professionals in a world of rapidly changing technology. This dialogue will guide the development of current and future education for library and information professionals.

Mobile technologies are changing the ways people live, work and play. They are significantly altering the nature of human interaction and the manner in which individuals and communities connect, communicate and use information. A small but growing number of libraries are beginning to apply mobile technologies to provide new services or to enhance traditional services. In doing so, these libraries are making themselves not only more available but also more relevant to their users. Recent discussion within the library profession has explored the trends and developments in mobile devices and their impact on library services. Interestingly, these discussions have not considered the impact these new devices will have or are having on librarians. What, if any, skills, knowledge and attitudes do librarians need to add to their existing toolkits to help them to design and deliver mobile services and collections? This chapter will fill the gap by providing the preliminary findings of a study aimed at exploring the changing skills and knowledge needed by the successful m-librarian.

Brief review of the literature

To date, there has been little discussion in the literature about the skills, knowledge and attitudes needed by librarians to operate successfully as m-librarians. The little literature that is available, however, indicates that m-librarianship represents a shift in practice that may require a different professional make-up for library staff. It is also useful to consider the literature related to skills and knowledge requirements for 'Librarian 2.0', as there are similarities between the changes heralded by both m-librarianship and Library 2.0.

Gavgani, Shokraneh and Shiramin (2011) did a content analysis of syllabi from Iranian universities offering a qualification in medical library and information science (LIS). The analysis aimed to identify instances within the syllabi of emerging trends and issues in LIS. The authors noted that the concept of m-libraries was absent from the syllabi and highlighted that this needs to be addressed. Gavgani, Shokraneh and Shiramin found that 'librarians' traditional skills and background knowledge are not sufficient to meet the changing needs of their customers. Librarians need to be empowered by new skills and information before going [on] to empower their patrons' (Gavgani, Shokraneh and Shiramin, 2011, 20).

Conversely, in a study designed to ascertain the skills, knowledge and attitude requirements for Librarian 2.0, Partridge, Lee and Munro (2009) found that the desired professional make-up was not dissimilar to that which librarians have always possessed. They found that Library 2.0 represented a 'paradigm shift' for the profession, requiring librarians to be agile and open to change. 'Librarian 2.0 requires a "different mindset or attitude". It is "challenging our mental models" and forcing us to think about and perceive our profession differently' (Partridge, Lee and Munro, 2009, 11). This aligns with the argument that Web 2.0 is 'an attitude not a technology' (Davis, 2005 in Secker, 2008, 217) and suggests that perhaps the paradigm shift is not about the acquisition of new skills and knowledge, but about inhabiting an attitude of openness, particularly where change is concerned.

The research

Semi-structured interviews were used for the data collection (Kvale, 2007). Six library professionals who were identified as being actively engaged in the provision of m-library services throughout Australia took part in the study. Participants' professional experience ranged from 3 to 36 years, with an average of 17.8 years; their ages ranged from 25 to 57 years, with an average age of 45.5 years. Only one male participated in the interviews. Almost all library sectors were represented, with individuals from academic, state, public and special libraries, although the academic library context dominated, with half of the participants coming from this setting. All participants were from metropolitan locations. The interviews were conducted in March 2011 and were recorded.

Findings

Although each interview tended to draw on specific themes of interest to that participant, there was also a great deal of common ground. Six key themes emerged in discussions of the skills and knowledge required by librarians in the use of mobile technology.

Technology

Not surprisingly, the role of IT or technology was discussed by all participants. Quite a number of the participants indicated that 'we are just buying [IT skills] in', one noting: 'that would be the equivalent of 20 years ago to writing your own library management system, you just wouldn't do that, so we're buying those skills in, outsourcing them'. A more pragmatic observation was offered by another participant: 'I don't think we'd be able to afford anyone who has those kinds of skills anyway, [we need to] concentrate on what we do well and find partners to deliver.' Although the majority agreed that librarians didn't need to know how to program or code, there was a feeling that some technical skills were required: 'We're doing content creation . . . and I suppose that involves some IT skills because we are creating videos, podcasts, screen shots . . . it's IT in the sense that we need to learn how to use the software in order to create those things'. Interestingly, two participants expressed uncertainty about the role of IT skills and knowledge. One noted that librarians 'would need good technical skills', but also concluded that 'it's more about library skills', about using the technology to 'make information accessible and to make the invisible visible'. The second also indicated that they were not sure 'whether it's the librarians [who] need to have all those skills or whether it's about being able to have that conversation with the person who's building it for you'.

Personal traits

Participants unanimously agreed that librarians need to possess a complex array of personal traits or attitudes. To be successful in the use of mobile technologies, librarians need to be adventurous, open to new things, willing to change and to keep learning. One noted: 'We need people who are happy to experiment and explore new ways of doing things . . . if we get too wedded to the old ways we become irrelevant.' Another noted the dangers of being 'too focused on ourselves and our own little world' and that we needed to actively work on not being 'insular in our attitudes and our views'. Librarians must also be willing to experiment, be creative and have imagination. One participant went so far as to suggest that 'imagination is the main thing, just imagining the way people are going to start getting information'. It was also acknowledged that librarians are not very good at knowing when to stop, and that to be successful with mobile technologies librarians need to have less fear of failure and less focus on perfection. As one participant observed, we need to develop the attitude of 'treating everything as beta that you either improve on or you stop and move on to something different'. Another participant observed that leadership and support from library management are needed in order to develop the personal traits and attitudes desired in library staff: 'We're not interfering at the management level and saying "don't do that, only do this", we're encouraging [them] to play and learn and open their minds up to the possibilities.' It was also noted that librarians need to role-model these desired qualities to other librarians, one participant observing that 'all it can take is a few people being one way or the other for the whole workplace to fall into the pattern'. The importance of personal traits and attitudes was strongly argued by another participant: 'If you've got someone who's talented and got the right attitude they'll explore and they'll learn quickly . . . every interview I have done in my life to get staff I've always made an excuse if I've found someone who's enthusiastic and talented and creative and imaginative. I don't really care if they are qualified or not.'

User focus

Many of the participants noted that mobile technology requires librarians to adopt a different relationship with their users. One remarked that it requires a 'move from [a] librarian knowing best to a librarian having a conversation with the community and hopefully being led by the community in terms of what we are doing'. Librarians need to be willing to work with their users as equal partners in the library service, they need to focus less on what the library has available and more on what the client wants to do. As one participant observed: 'They see technology and they imagine what the potential is . . . and we go and talk to them and say "we're putting this in . . . come up [with] some concepts for us" and they come up with some concepts for us.' Another participant even noted that they were proactively endeavouring to be 'more like the students'. As noted earlier, although library clients or users are starting to engage with mobile technology, not all are as fluent with it as might be expected; consequently, the 'librarian as teacher' is still important: 'They come in familiar with iTunes and they might be familiar with Twitter or Tumblr but they are not familiar with other things they could use . . . so we've moved into that space of teaching them things that they want to use, and they should know how to use because they are fairly intuitive but they aren't exposed to for whatever reason. Information literacy classes are not just about unfriendly databases.'

Communication

Communication was identified as essential for librarians to succeed in using mobile technologies within their services. In this instance communication was more than just the ability to engage in written and oral discourse in diverse formats and media; it also included an array of more complex dimensions and aspects. Librarians need to know how to be advocates and lobbyists, one participant observing: 'It's going to require a lot of campaigning both to bring our users to the table and to work together with other libraries to try and bring publishers to the table.' Librarians will need to be good at negotiation and influencing: 'It is unlikely that a lot of libraries are ever going to have the money to do these things on their own and so they may need to partner with other organizations or perhaps lead the larger organization down that path if they feel that it's important to their users.'

Collaboration

Almost all the participants acknowledged that collaboration was a core element for success, especially with IT departments and managers: 'The IT people are very aware of new technologies that we can apply . . . so they're really good at bringing some of that forward, a lot of the impetus for change has come from the IT people.' This point was raised because it was acknowledged that 'we can't do everything and we have to call on our community and our network of people to use everyone's skills'. Librarians need not only to appreciate the need for collaboration and for embracing different perspectives, they must take action to ensure that this happens. One participant noted the need not to be reliant upon the IT services offered by their government organization, which could not always provide the degree of support required, but to establish an in-house IT team: 'They're part of the library, they form part of the team, they're on every project and just about everything we do is technically based or has a technical side to it.'

Research and development

Research skills were seen by some participants as essential for librarians developing services with mobile technology. This point was raised because it was acknowledged that 'everyone is feeling their way through this change' and that consequently it involves a considerable degree of problem solving and critical thinking. One participant noted that 'there's been a lot of jumping on bandwagons with new technology and really you need to critically evaluate what's going to help you and whether you are justified in spending the dollars'. Research is not always about finding a solution, and one participant remarked that they frequently undertook research simply out of a desire to understand more about the changes taking place around them. Research provides an opportunity for librarians to 'step back and take a look at the bigger picture and think a bit more strategically about what direction you want your library to go for the next period'. However, participants also acknowledged that some libraries will find it harder to engage in research and development than others, given limited resources and/or organizational support; although one participant observed that 'the industry are just so hopeless at doing much research and development themselves . . . they don't really do research anymore so they're not even aware of what's happening within their own industry'.

Discussion

But haven't librarians always been required to have these skills and knowledge? Interestingly, while most participants responded to this question in the affirmative, there were differing degrees of agreement. Some participants agreed completely that the skills and knowledge being discussed were not new; others felt there were a few new IT skills emerging, such as content creation. Another view was that what we need is a mash-up of skills and that libraries may need to reconsider their staffing mix and acknowledge the need to involve more non-librarians within the suite of personnel, who can offer new and alternative perspectives, skills and knowledge: 'I think we should have a mixture of staff to see the blending of more of those skills based on . . . the individual talents of the people.' However, one participant feared that this might result in '[trapping] them in the library, then they become more like us and less like them . . . I don't know, I am not sure how to do this yet'.

Conclusion

Although the scale of this project was small, it was a valuable exercise in developing an understanding of the skills, knowledge and attitudes required by librarians to operate successfully in the m-libraries space. It revealed interesting synergies between the attitudinal requirements for the m-librarian and Librarian 2.0 (Partridge, Lee and Munro, 2009) and confirmed our impression of the skills and knowledge needed by m-librarians. Information professionals still need the skills that have traditionally been associated with their profession, but in this new world of hyperconnectivity and mobile computing it is our attitudes as professionals that need to shift, so as to embrace emerging technology and the possibilities it offers. Success in the domain of m-librarianship is not so much related to the possession of high-level IT skills as about taking our existing skills and applying them in this new environment and, importantly, about openness to the redeployment of our skills. The challenge for LIS educators, then, is to consider how we might provide learning opportunities that allow our students to develop the right attitude for success in this rapidly changing information landscape. But perhaps that challenge is nothing new, either.

References

Gavgani, V. Z., Shokraneh, F. and Shiramin, A. R. (2011) Need for Content Reengineering of the Medical Library and Information Science Curriculum in Iran, *Library Philosophy and Practice*, 477, January 2011, 8.

Kvale, S. (2007) Doing interviews. In Flick, U. (ed.), *The Sage Qualitative Research Kit*, Sage.

Partridge, H., Lee, J. and Munro, C. (2009) *The Skills and Knowledge of Librarian 2.0: preliminary findings*, Queensland University of Technology and Australian Learning and Teaching Council.

Secker, J. (2008) Social Software and Libraries: a literature review from the LASSIE project, *Program: electronic library and information systems*, 42 (3), 215–31.

11

There's a librarian in my pocket: mobile information literacy at UTS Library

Sophie McDonald

Introduction

At the University of Technology, Sydney (UTS), Library we are considering how mobile and social technologies are changing the way our clients engage with information. In everyday life, at home and university, our clients are engaged in problem-based learning, collaboration and content creation. They are increasingly relying on mobile devices, flexible delivery of services and 24/7 access. As information literacy instructors, we recognize the need to be proactive in meeting the changing needs of clients and have been experimenting with ways of supporting learning in this new environment. This experimentation has led to the development of a more game-based learning model so as to move our information literacy programme beyond the classroom and into flexible, personalizable, 24/7, physical, digital and mobile spaces.

In 2009, UTS Library launched a mobile-friendly website accessible from most mobile devices. The IT team decided not to create an app because it wanted to be platform independent, and it didn't create a separate mobile site because it didn't want to manage content in multiple places. Instead, it developed a custom template engine that detects which device is requesting the website and returns the content in the appropriate format for the device. The team also wanted to avoid an all-or-nothing approach and decided to go with a rolling series of releases rather than wait for the entire site to go live. Clients can access a range of high-use content and search for resources using the mobile-friendly catalogue. When an item is found in the catalogue you can see a location map showing where to find it. The Library website is being made progressively mobile friendly and room bookings have recently been added. By the end of 2011 the whole website should be available in mobile format. Currently about 6% of our website traffic comes from a mobile device and 85% of these are Apple devices. We expect traffic to our mobile website to increase dramatically over the coming year as more and more of our clients upgrade to the latest smartphones and tablets – and based on current trends, these may not remain predominantly Apple.

Making it social

Social media sites are increasingly being accessed on mobile devices and it is essential to have a presence in these spaces so that we can connect with clients in their online communities. In 2010 we expanded our existing social media presence from Facebook, Twitter and YouTube to include Foursquare, Flickr, Diigo and two blogs. Foursquare is a social and mobile geo-location game that we use in our strategy to promote general information literacy, encounter clients in their online 'spaces' and encourage peer-to-peer promotion of the Library.

Our library news automatically feeds through to Facebook and Twitter but it is important to have personal updates, so library staff share links to online tutorials, workshops, study tips, Flickr photos and competitions. We have lots of 'likes' and comments, but, more importantly, we can see how many people click through to our website from Facebook and Twitter. In our recent monthly statistics we had over 600 monthly active users on Facebook, out of about 1000.

We run promotions and competitions to engage with our clients, promote digital literacies and encourage co-creation of content. For example, we started our YouTube channel by running a competition for students to make a video about what makes UTS Library so great. Our winning entry now has over 2000 views and our channel overall has over 35,000. In 2010 the competition was in digital storytelling and in 2011 we are running a data mash-up competition.

We use Flickr to share our visual history and allow our community to share photos with the UTS Library group. We have two blogs: one for researchers and one for our reading club. Our researcher blog also links to a Diigo group for collaborative bookmarking and resource sharing. These are all ways of engaging with our community outside the physical Library and the official Library website. We are sharing with our clients on their turf and learning by doing.

Connecting the physical and digital

In 2009 we started to consider how we could deliver mobile content to our clients to support learning, teaching and research. We used Quick Response (QR) codes as a way to provide contextual self-help at the point of need for just-in-time learning. For example, a QR code on our self-check-out machine links to a short video on how to use the machine. We also use QR codes on our promotional posters, bookmarks and brochures, linking to our website and instructional videos on our YouTube channel that can be viewed on a mobile device. Like many libraries, we have found that take-up of QR codes is low, but this has recently spiked with the addition of our 'flush ads', which are promotional posters on the backs of toilet doors. Despite the low take-up, the trend of scanning physical things to acquire digital information is expected to grow, and so we will continue with QR codes while the technology develops.

Making it fun

In 2010 we started to develop a more game-based learning model for delivering information literacy, and QR codes provided an opportunity for this during our first Library Fun Day. Fun Day is held during orientation week, with the aim of introducing new students to the Library in a fun and friendly way using games, prizes, goody bags and free food.

Library Fun Day expanded in 2011 to include a 'technology petting zoo' for clients and staff to touch and feel a collection of e-book readers, iPads and other mobile devices. The area was very popular throughout the day, with many people trying the devices for the first time and those who were more experienced sharing their knowledge with others.

One of the Fun Day games is a Library Treasure Hunt where participants collect QR code clues as they race around the Library undertaking a series of activities and trying to be the first back to claim the treasure (Figure 11.1). In just 15 minutes most are able to navigate all of the activities and cover everything normally taught in a one-hour introductory information literacy class! Feedback from participants is positive and the winners receive movie or iTunes vouchers. We use paper clues as a backup for anyone not able to scan the QR code clues, and at Fun Day 2011 there was a big increase in the number of students able to scan the QR code clues.

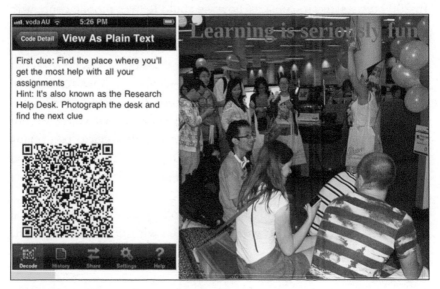

Figure 11.1 QR code Treasure Hunt

The Treasure Hunt was such a success that it is now being used as a model for many of our first-year information literacy sessions, most notably, for the last two years, for several hundred first-year nursing students. These students spend much of their time away from the University on practice placements and often require follow-up assistance at the Research Help Desk, even after attending traditional information literacy classes. Library and faculty staff acknowledged their need and collaborated to develop this new way of delivering information literacy skills. Data is being collected on these sessions via pre- and post-testing and is yet to be analysed; however, lower numbers of students returning to the Research Help Desk indicate successful learning outcomes.

Making it visual

We are moving away from a text-based model for delivering online information literacy because we recognize that many of our clients are visual learners. We aim to meet their expectations by developing more dynamic and interactive online tutorials for acquiring traditional and non-traditional information literacy skills. For 12 months information services librarians have been creating content specifically for the mobile environment by creating screencasts and vodcasts using Camtasia, iMovie, jingPro and Adobe Captivate. We are teaching ourselves how to play with these tools so that we can come up with low-cost online tutorials that support learning, teaching and research. We upload these to YouTube so that they can be accessed on mobile devices and embedded in our website or learning management system. We have recently made three in a series for researchers called Make Me Famous, which demonstrate how to publish more strategically and prove the impact of your research by using tools such as Thompson's Journal Citation Reports and the h-index.

Developing mobile literacies

Mobile devices will be the main tool for accessing information in the near future and it is important to support our clients to use these tools effectively. In an attempt to develop mobile literacies, in 2010 we started running mobile searching workshops to increase awareness of our mobile website and increase usage of our mobile services. Library staff run the workshops and help clients to connect to the free university Wi-Fi network, search our mobile Library catalogue, get a location map of the item they are looking for, search mobile databases and save PDFs for reading offline later. We are testing mobile databases as they become available to check that they can operate through our institutional login, which is not always possible with apps. In the workshops, we also discuss apps that our clients can use to help with managing information, like Dropbox, Diigo and Stanza.

In 2011 we have seen a large increase in the number of students with the latest smartphones and no problems connecting to Wi-Fi. While the mobile searching workshops are obviously a learning opportunity and fit the traditional notion of information literacy, other things we that are doing are developing mobile literacies in more subtle or non-traditional ways. Examples are the technology petting zoo, competitions and use of QR codes, all described above.

Learning by doing

To achieve all of these things, UTS Library staff are learning from each other and learning by doing. We are being encouraged by senior management to play with mobile devices in order to imagine both how we can use the technology ourselves and how we can support our clients using the technology. To help in this, the Library purchased iPod Touches and iPads for staff to use for roving reference and a selection of e-readers, including iPads, for staff and clients to borrow and play with. The IT department is aiming to make more of our work practices mobile friendly and recently trialled a device that attaches to an iPod Touch for mobile inventory management. As more and more of our clients are using mobile devices to support their learning, teaching and research it is important that we understand the technology and how our clients are using it so that we can start to develop new mobile services.

Conclusion – the future is mobile

One of the reasons for our developing mobile literacies and promoting mobile services is that we are currently planning our Library of the Future, which will mean not only a new library building but also a new approach to using space and delivering services.

We are putting over two-thirds of the physical collection into an underground automated retrieval system. This means that we need to develop new ways of allowing online browsing and serendipitous discovery by harnessing crowd sourcing and personalization. For example, a client can search for a book on their mobile device on the way to the library and the system will have it ready for collection when they arrive!

Along with the retrieval system, we are implementing radio-frequency identification (RFID) technology and researching ways that it could help to create a 'smarter library'. There are possibilities for the whole library experience to be personalized using RFID smart-card technology. This is an emerging area for libraries and we plan on experimenting and playing with the possibilities of mobile technology to deliver services and support to our clients so as to meet their learning, teaching and research needs. Moving further into the future, we will be looking for opportunities to combine our mobile and RFID technologies with augmented reality so as to enable greater opportunities for harnessing social media, mobile spaces and gaming within our information literacy programme.

The projects described in this paper are the work of a dedicated team of librarians and IT staff who are working together to play with mobile technologies and develop services that are flexible enough to change with the technology. Many of the projects are in their initial stages and will develop further as we learn from each other and our clients. We know that mobile technology will play a strong role in the library of the future, but exactly how is not yet evident. However, if we continue to play with the technology we will be better able to keep up with it as it changes and deliver innovative mobile services.

12

Exploring student engagement with mobile technologies

Julie Cartwright, Sally Cummings, Bernadette Royal, Michelle Turner and JoAnne Witt

Introduction

Smartphones, in essence mini wireless-enabled computers, have begun to outsell personal computers in many countries (Little, 2011, 267). Certainly at Charles Darwin University (CDU) Library we have observed the increasing presence of both smartphones and other mobile devices such as iPads and tablets. Greenall (2010, 16) urges libraries to engage with this 'user-driven technology' trend or risk being out of step with their clients.

But what should we be doing? Do we need a mobile site? Or some apps? And what about Quick Response (QR) codes, those two-dimensional barcodes that are popping up in a variety of places? Some worry that QR codes cannot live up to the hype (Carlucci Thomas, 2010, 34), or that the barriers to their use are too high (EDUCAUSE, 2009). However, the University of Huddersfield and University of Bath successfully adopted QR codes in 2009 (Walsh, 2009), using them to link clients directly to a range of reference desk and help services, including micro instruction tutorials (Walsh, 2009).

We, as librarians, need to learn more about mobile devices. We need to explore, play and have fun investigating the possibilities. Indeed, far from heralding 'the end of usefulness for libraries or librarians', Hahn (2008, 284) suggests that providing digital library content on mobile devices is a 'new existence for the Library'. As mobile technology usage increases, librarians need to be able to develop an understanding of trends in mobile use. This knowledge needs to be incorporated into our skill set. We, at CDU, need to gather the evidence required to make appropriate strategic decisions in light of the fact that our library is a small institution and our funding is limited. We cannot afford to waste staff time and resources on products our clients do not want.

Challenges and issues
How do we engage with students?

Charles Darwin University is a dual-sector institution formed in 2003. The student cohort includes Year 11 and 12 students, Vocational Education and Training (VET) in Schools programme Certificate I to IV VET students, as well as higher education (HE) and postgraduate students. Our students are drawn from the local area, across Australia and overseas and comprise indigenous and international students, school leavers and mature students returning to study. Each of the three campuses, Casuarina, Palmerston and Alice Springs, has a library.

Courses and units may be delivered online, at any of the campus locations or, often, in remote areas. Of the approximately 6300 HE students enrolled at CDU in 2011, two-thirds are fully external, and a further 600 students are enrolled in mixed mode. The remainder are enrolled as internal students.

Our prime objective in developing this project was to investigate our patrons' knowledge and use of mobile technologies. We chose to focus on QR codes because they are simple and cheap to produce, they are specific to smartphones and we had observed some good examples of their use within academic libraries, such as at the University of Queensland and the University of Technology, Sydney, in Australia, and the University of Bath and the University of Huddersfield in the UK. To achieve our objectives as well as promote awareness of the range of library services and resources within a fun atmosphere and raise the library's profile, we decided on a library Treasure Hunt activity. We wanted the activity to be inclusive and to reach out to our high percentage of external students. Various prizes were offered as an incentive for students to participate.

DIY approach

The Liaison Librarian team at Casuarina Campus has five staff members and all tasks were shared. We read up on QR codes, QR code readers and practical ways of using QR codes. As the University is small, our budget for the Treasure Hunt was extremely limited. Our project was completely reliant upon staff time, enthusiasm and creative thinking.

Throughout the planning of the activity we kept in mind that we wanted students to have fun while they were learning, and also that QR codes are a great way of linking the physical and virtual environments. We took advantage of some free tools available online to create a variety of content for the activity. The Liaison Librarians were encouraged to play and to up-skill themselves. We used Animoto to create a video clip on using the photocopy and print room printers. We also created video clips using Blabberize to provide clues. Animal mascots used in the activity were created using Creative Commons-licensed photographs from Flickr, which were converted into cartoons using BeFunky.

One of the key features of this project was making it measurable and producing results that we could analyse meaningfully. To that end we built in measures right from the beginning. We created a bit.ly account, which we used to produce shortened URLs for all the links in the Treasure Hunt. Bit.ly collects statistics on how often and when links are clicked on, as well as information about users' locations and referrers. QR codes are also created by bit.ly for each shortened URL, and these were used in the Treasure Hunt and generated statistics.

Promotion was an important aspect of the activity. A Treasure Hunt LibGuide was created containing information about the Treasure Hunt, QR codes and links to downloadable QR code readers. We tweeted about it, notices were posted on the University learning management system and University e-mail lists. We purchased cheap white T-shirts, created our own QR-code transfers and ironed them onto the backs of the T-shirts. Badges (made using reusable plastic name tags) urged our clients to 'Scan me now, Ask me how'. Library staff were encouraged to wear these, helping to generate and maintain enthusiasm amongst front-of-house staff. Posters were designed and printed in bright colours and placed throughout the University campus, and PowerPoint slides were created for the Library's digital display screen located in the Library foyer.

Sponsorship was sought for Treasure Hunt prizes. This was most successful among businesses that had a direct relationship with the University and students, such as the CDU and Dymocks bookshops, the on-campus coffee shop and Corporate Express, a national office supplies store.

Treasure Hunt

Four Treasure Hunt pathways were created for the various student groups: on campus (at the Casuarina Campus Library) students with and without smartphones; and online students with and without smartphones. To maintain a playful atmosphere and simplify navigation around the physical library, as well as around the Library website, each of the groups was allocated a different animal mascot to follow during the activity. The Treasure Hunt involved visiting a range of locations throughout the Library, both physical and virtual, gathering clues and secret words. Some clues linked to short, fun video clips. Each group had its own secret phrase so that we could differentiate what mode students were using to complete the Treasure Hunt. CDU students at campuses other than Casuarina were encouraged to participate online. Those participating online were guided around the Library website.

At Casuarina Campus, on-campus student groups began by collecting a postcard from the Library circulation desk that provided them with their initial clue and was used to record their secret phrase. Scheduled times were offered for students to come along and participate in the Treasure Hunt guided by a Liaison Librarian. Class groups participated during scheduled Library workshop sessions.

Findings

The Treasure Hunt ran for just over four weeks and 223 students participated. Of the entries, 153 (69%) were online and 70 (31%) were on campus. For both groups, we found that approximately 10% used QR codes. We also collected statistics from the Treasure Hunt LibGuide, which was the activity's starting-point and information page. There were 1205 visits to the LibGuide in February and March, as well as 203 visits to the About QR Codes page, 73 clicks on links to download QR code readers (28 of the 73 were for a QR code reader that can be used on PCs) and 30 visits to the Sponsor Information page.

Our bit.ly statistics showed that 82% of visits to the Treasure Hunt LibGuide via bit.ly were from the QR code that appeared on our promotional material, such as T-shirts and posters. Four per cent of clicks were via the bit.ly link that was tweeted on our @CDUniLibrary Twitter account. We also attempted to measure clicks on QR codes created for text-only messages within the Treasure Hunt, but unfortunately they did not record successfully.

Survey

Along with the announcement of the prize winners we sent out a brief survey using SurveyMonkey, the online survey tool. We received only 15 responses. This, and the fact that many prizes remained uncollected, confirmed our belief that e-mails to student e-mail accounts are not necessarily effective for reaching students. Of the 15 students who responded, roughly half had first heard about the Treasure Hunt through the CDU website. Most of the others had learned about it from Library staff or from their lecturers. Eighty per cent had found it easy to follow, roughly half said it was fun and almost as many found it informative.

We also wanted to know a bit more about our students and how they use mobile technology. A third of respondents stated that they owned smartphones (we provided a description of smartphone brands to avoid any terminology issues). Of the remainder, one planned to purchase a smartphone, but the rest (60%) didn't own one.

For half of the respondents, the Library Treasure Hunt was the first time they had seen a QR code. A quarter said that they had known about QR codes beforehand; another quarter still didn't know what a QR code was. In terms of future usage, about a quarter of respondents expressed interest in using QR codes.

Beyond QR codes, we asked students whether they would use their mobile devices to access Library services. Each of the services we suggested (such as catalogue and database searching, online tutorials and contact information) received a positive response from about a quarter of respondents, while 20% said they weren't interested and 40% answered that it was not applicable at this point.

During this period the Liaison Librarians informally recorded their observations of smartphone usage in the library. Uses included students saving catalogue records on their phones in order to make enquiries or to locate the item on the shelves. Observations made during headcounts were that many students now carry smartphones.

What worked

- Our planning was extremely thorough – we thought through all the possible scenarios, walked through the activity with the help of testers and built-in measures from the start.
- We organized for class groups to use the Treasure Hunt as a learning activity. Some groups came to the Library, while others completed it online in their computer-enabled classrooms.
- The Treasure Hunt LibGuide worked well as an information tool for QR codes as well as for the online starting point.
- Multimedia and video elements were well received and allowed us to inject a sense of fun and light-heartedness into the activity.
- Communication and co-operation with other library staff created buy-in and encouraged all staff to promote and support the activity.

What didn't work

- Treasure Hunts held at scheduled times were poorly attended – it seems that students preferred to do the activity at times that suited them, and under their own steam and mostly online.
- CDU student e-mails were ineffective, both as a promotional tool and for contacting prize winners – most of the coffee vouchers and one of the larger prizes remain uncollected.
- When taking large class groups through the Library it was not practical to get those students with a smartphone to download a QR code reader before doing the Treasure Hunt.

What we learned

- Students undertake activities like this for the novelty and the fun as much as for the prizes (if not more).
- Ten per cent of participants were prepared to use QR codes and many more were interested in how they work, but it appears that most did not feel compelled to download a reader.
- The online version of the Treasure Hunt had the higher participation rate, which is consistent with our external student cohort and could also indicate a preference for online activities.
- There is a wealth of useful, easy-to-use freeware out there to be tapped into by libraries with small budgets.

The Treasure Hunt enabled us to collaborate effectively with other sections of the library and to co-ordinate a very successful orientation programme.

While we were pleased with the level of participation that we achieved, it was difficult to capture students' attention, both on campus and online. In future we may try more direct promotion through the lecturers and more classroom sessions.

Conclusion

We are seeing increasing levels of mobile phone and smartphone use in the Library. QR codes can be used in a wide variety of activities, both informational and promotional, and their use can be measured.

Our results indicate that the potential for further incorporation of QR codes into Library services is worth pursuing. Due to our limited resources, we will focus on projects that have a high impact for low cost. We believe that QR codes are most effective for linking the physical and digital worlds. We plan to use them initially to link to instructional video clips which can be viewed on mobile phones. These micro instructional tutorials are well received, and easy to create using free online tools. For example, we plan to put a QR code poster on the wall in the print/photocopy room linking to a video tutorial on how to add money to print cards. Putting QR codes into our promotional material, such as the LCD screen in the Library foyer and posters, will allow students to link to more information about Library events and services. In the longer term we will monitor QR code usage and, if it is warranted, investigate whether QR codes can and should be incorporated into our catalogue records.

Finally, smaller libraries with tight staffing and budgetary resources can create engaging activities by using free online tools, encouraging play and making the most of existing staff knowledge and enthusiasm for building fun learning activities.

Selected web tools

Animoto, http://animoto.com
BeQRious, www.beqrious.com

BeFunky, www.befunky.com
bit.ly, http://bit.ly
Blabberize, http://blabberize.com

References

Carlucci Thomas, L. (2010) Gone Mobile? Mobile catalogs, SMS reference, and QR codes are on the rise – how are libraries adapting to mobile culture? *Library Journal*, 135 (17), 30–4.

EDUCAUSE (2009) 7 *Things You Should Know about QR Codes*, EDUCAUSE Learning Initiative, http://net.educause.edu/ir/library/pdf/ELI7046.pdf.

Greenall, R. T. (2010) Mobiles in Libraries, *Online* (March/April), 16–19.

Hahn, J. (2008) Mobile Learning for the Twenty-first Century Librarian, *Reference Services Review*, 36 (3), 272–88.

Little, G. (2011) Keeping Moving: smart phone and mobile technologies in the academic library, *The Journal of Academic Librarianship*, 37 (3), 267–69.

Walsh, A. (2009) Quick Response Codes and Libraries, *Library Hi Tech News,* 26 (5/6), 7–9.

13

It's just not the same: mobile information literacy

Andrew Walsh and Peter Godwin

Introduction

Libraries are increasingly developing services that take into account the massive impact of mobile devices on their users. This chapter looks at what this might mean for information literacy.

Some databases and library catalogues have mobile versions. Growing numbers of our users have the devices to access these. When a library user is as likely to search for information on a mobile phone, tablet or hand-held gaming device as on a fixed PC in a physical library, this could change the concept of what it means to be information literate. How will mobile search change the discovery, evaluation and reuse of information? We will first consider whether we need to develop a new lens through which to view the literacies required by our clients. From these theoretical considerations we will then proceed to a quick review of experience of mobile devices gained at two UK new universities: Bedfordshire and Huddersfield.

Libraries could presently be in their greatest period of change: we have moved on from the hybrid library, we are facing a public perception of electronic delivery of all content, and Web 2.0 has given everyone the potential to author and to share. Now add to this the phenomenal spread of mobile devices the world over. Information is becoming mobile and social. It also appears that the economic recession that has affected large parts of the world may not slow these developments. Librarians must spend more time over coming months conditioning their services to the mobile environment.

How does mobile technology affect information literacy?

The image of young students who are continually wired into the online environment, texting and multi-tasking, has been overdone. It has been easy to exaggerate the differences of the web generation. The CIBER reports (Rowlands et al., 2008) have done much to debunk the hype that used to surround them. The truth is that we have all become part of the web generation, to a greater or lesser extent. Consider mobile devices: their adoption is across the whole of society. But what will it mean when searching is done on the move and with an expectation of immediate access to information? It is likely to involve an extension of existing trends whereby search is carried out at speed, with lack of reflection on the results and less reading of the actual content. Also, there will be a requirement to join these devices up to existing personal learning environments and to new, developing platforms. Because we are already seeing a merging of literacies (ITC, information, media, visual, etc.) the management, manipulation and reformulation of all this content is being seen by some as 'transliteracy'. The insertion of mobile devices into this equation simply adds a new twist.

But what is the nature of that twist? Do we know how information seeking and usage changes on mobile devices? At the moment the evidence is limited, but the current literature does provide evidence for four areas in which 'mobile information literacy' may vary from 'fixed information literacy'.

Where it is manifested

Traditionally, searching for information, evaluating it and using it were expected to happen in a limited range of contexts. Searching for information may happen in a library, from a fixed workspace, possibly at a fixed multi-purpose computer with a large screen. Mobile searching can happen in any place that has a mobile phone or wireless internet signal, and from a range of devices with massive variations in functionality. Searching no longer happens in fixed, controlled environments, but in random, messy, uncontrolled ones (see Table 13.1).

Church and Smyth (2008), in a diary study of mobile information needs, found that over 67% of their participants' information needs were generated while the user was mobile. The quantity and penetration of mobile internet-capable devices means that people can

Table 13.1 Searching in fixed and mobile environments

	'Fixed' information literacy	'Mobile' information literacy
Where?	Largely in 'set' places. At a desktop computer (with little variation in software); at a fixed workplace; within a library.	Anywhere; any mobile device (phone, games device, e-book reader – massive variations in device functionality).
What?	Anything and everything.	Normally quick information, often context or location specific.
How?	Range of established tools to access and manage wide range of information sources. Standard search engines.	Often narrow apps and individual specialist sites rather than open web.
Time spent	Varies. Often slow, long access. People spending long periods searching for, organizing and extracting information, especially for academic use.	Quick/fast only. Shorter searches. Little pondering and extracting of information. Favours short chunks of information. 'Convenience' of device.

increasingly attempt to meet their information needs as they occur. Indeed, Heimonen (2009) found that amongst already active mobile internet users, virtually all 'on the move' information needs were addressed through mobile devices, as they occurred, with the only failure to address a need being caused by a mobile phone battery running out!

What searches are carried out

Mobile information needs are dominated by the desire for quick access to often context-specific information, particularly regarding local services, travel and trivia (Church and Smyth, 2009; Heimonen, 2009). Whereas when searching and using information in a fixed, traditional location we may search for anything and everything, this isn't the case for mobile use. The searching we carry out on a mobile device is much more likely to be an additional activity rather than the sole focus of our attention, and will therefore be influenced by the primary activity in which we are also engaged – that is, the context in which we find ourselves (Hinze, Chang and Nichols, 2010).

The information we seek on the move is about facts, and small elements of information. There is likely to be limited evaluation of the information we find, and little opportunity to take away detailed information and derive new knowledge from it.

How we search

Searching for information online from a desktop computer allows access to a wide range of established tools and information sources. It is normal to start searching for information with a generic search engine, and this may lead on to more specialist sites or search tools. Such searching could be characterized by the breadth of sources and tools available and used. Mobile searching, however, is heavily influenced by the natural constraints of using a device with a small screen and a small or virtual keyboard, and may be characterized by the narrowness of the sources used.

Rather than search the open web, smartphone users are also tending towards the installation and use of specialist apps.

Time spent on searching

Heimonen's (2009) study found that 35% of information needs occurred in the home. Even though a fixed computer (or laptop) might have been available, the speed, proximity and convenience of using a mobile device trumped the more powerful device. People turn to their mobile devices for quick and dirty searches for information. They want to know something, and they want to know it now!

A typical comment from some recent interviews carried out by one of the authors is: 'I just love the thought of not being tethered to go and fire up the old laptop or desktop machine . . .' The interviewee tended towards using a mobile device for speed and convenience.

University of Bedfordshire

The University of Bedfordshire is a new university with libraries on two major campuses (Luton and Bedford) and several smaller units serving 23,000 students. Recruitment of international students is particularly strong, and we will focus on the views of these students later in the chapter. A small general survey during summer 2010, together with the obvious prevalence of mobile devices among students, prompted the development of mobile-friendly access to library resources. Mobile devices also seemed to provide a way of connecting with library users to aid the development of information literacy.

Quick Response (QR) codes offered the most promising way of helping to exploit our stock and also provided a means to explain procedures or how to use unfathomable equipment on site. We had already developed a series of short library videos that we called 'Just a Minute', covering anything from a basic introduction to the catalogue to print credit machines, self-return machines and referencing. These were uploaded to the Library website (http://lrweb.beds.ac.uk/libraryservices/whoweare/videos).

The possibility that QR codes offered for posting links to these videos at the point of use was novel and exciting. For example, a QR code was used for a link to a video on how to operate the new movable stack for our journal collection at the Luton site. We have experimented with the use of QR codes in subject leaflets, to encourage connection with our best resources. Certain textbooks have always been in exceptional demand, and although they might sometimes be available as an e-book, our users still opted for a print book. We have therefore been using QR codes on the bay ends of bookshelves to promote the e-versions of our most popular titles (Figure 13.1 overleaf).

During the last year we have developed in-house a free learning resources application for Android phones that links to the catalogue, our website, a list of our 'Just a Minute' videos, and a GPS 'Find a Campus' guide to the local campus; it e-mails through the auto-enquiry function or calls us directly to renew items. (See http://lrweb.beds.ac.uk/libraryservices/whoweare/apps/android.)

However, we still needed a direct library catalogue app, and for this we chose the commercial LibraryAnywhere software, which is compatible with iPhone, Android and Blackberry devices. (See http://lrweb.beds.ac.uk/libraryservices/whoweare/apps.)

Finally, we have adopted the commercial Z-Bar software for both iPhone and Android devices, which allows the scanning of barcodes (importantly, of book barcodes in shops) to match against the Library catalogue.

Figure 13.1 QR code used to promote the e-version of a title in heavy demand

Impact

So far, we have listed our initiatives but not provided any indication of their impact. The rate of download of the applications has been encouraging. Take-up of QR codes was likely to be slow: barriers were the students' being unaware of them, the need to download a free application like Beetag to their devices and the need to log into the wireless network on campus. Promotion was needed, and championship by academic librarians in their encounters with students.

In April 2011 we decided to take a snapshot to test the ownership of devices and knowledge and take-up of our mobile services with a major client group: a large cohort of MBA students, the vast majority of whom were from the Indian subcontinent. Responses were gathered from 172 students (about 10% of the student cohort), of whom 116 considered they were using a mobile device as a learning tool, but only 25 admitted to having an iPhone, 24 a Blackberry, 20 an Android, 3 an iPad, and 2 an e-book reader. Only 27 said that they knew of our Android application for library services and a mere 10 had used it; 36 knew about QR codes and 21 were using them. Many of these students had previously seen a demonstration of a QR code. Despite some clear faults in the survey design, the message was that these students were not so advanced in mobile use as one might have supposed. Familiarity with and use of these devices will be crucial in a business environment, so we feel encouraged to be much more proactive with future cohorts. This can be done most effectively by following up the survey by holding small focus groups and interviews with members of this client group.

University of Huddersfield

The University of Huddersfield has in the region of 24,000 students, with representation from over 130 countries. It is a post-1992 university with a rich history of vocational education reaching down to the present, and is among the UK's top ten providers of 'sandwich courses', during which students undertake a paid work placement in industry or commerce. In recent years we have noticed increasing numbers of mobile devices being used in the Library, with a current trend towards smartphones rather than standard 'feature phones'. Indeed, a University survey in 2010 found that roughly three-quarters of our students had a mobile phone contract that allowed access to the internet, the majority of which had effectively unlimited data. We have therefore experimented with several services that take advantage of students' own mobile devices, and a few of these are outlined below.

Although text messaging (SMS) has been in existence for many years, with a high percentage of people in the UK having access to it, until recently we had not taken advantage of it. During the last couple of years we have tried various text messaging services, the first being a simple 'text a librarian' service that allowed members of the University to text questions through to a librarian, alongside the many other ways of contacting us. More recently, this simple text service has been taken up by others, our most popular mobile service currently being the ability to text NOISE to a number, along with a location, to report inappropriate behaviour in the library.

Getting more directly involved with information literacy and text messaging, we produced a set of ten text messages to support our face-to-face induction process. These were sent out during the first term of the academic year to remind people of useful information that we traditionally cover in the standard induction process.

We have also used students' mobiles in classes, experimenting with different ways to use them for polling, similar to the interactive handsets or clickers that many educational institutions use. Currently, we are tending to use Poll Everywhere (www.polleverywhere.com), which allows similar functionality to clickers, but using the students' mobiles. They can vote via text message, Twitter or the web. The basic service is currently free for class sizes up to 30.

QR codes have been used in many ways, particularly in printed hand-outs to link the plain physical paper to extra videos, quizzes and other online materials, thus turning them into interactive hand-outs. These seem to be a great way of bringing additional information literacy materials to our users, but can only be an 'extra' at present because so many people either cannot, or are not willing to, access QR codes.

Some of the materials linked to from QR codes are videos that we have tried to make available in a mobile-friendly format. We have settled primarily on uploading videos to YouTube to facilitate this, rather than trying to provide a range of alternative formats.

We suspect that mobile search will be increasingly important, and our resource discovery service (Summon), which includes the Library catalogue, works well on mobile devices.

Conclusion

It is likely that mobile devices will continue to make a difference to the ways in which people interact with information – a difference that is likely to accelerate. We do not yet know how mobile information literacy may be different from traditional views of information literacy, although clues are already emerging.

We are providing a range of services, only some of which are outlined above, to help to develop information literacy within our universities. In the near future we need to be more aware of how information literacy itself is changed by mobile devices so that we can continue to provide appropriate and useful services to staff and students within our institutions.

References

Church, K. and Smyth, B. (2008) Understanding Mobile Information Needs. In
  Proceedings of the 10th International Conference on Human–Computer Interaction with

Mobile Devices and Services (MobileHCI '08), ACM.

Heimonen, T. (2009) Information Needs and Practices of Active Mobile Internet Users. In *Mobility '09: Proceedings of the 6th International Conference on Mobile Technology, Application & Systems, held in Nice, France*, ACM.

Hinze, A. M., Chang, C. and Nichols, D. M. (2010), *Contextual Queries and Situated Information Needs for Mobile Users*, Working Paper 01/2010, University of Waikato Department of Computer Science.

Rowlands, I., Nicholas, D., Williams, P., Huntington, P., Fieldhouse, M., Gunter, B., Withey, R., Jamali, H. R., Dobrowolski, T. and Tenopir, C. (2008) The Google Generation: the information behaviour of the researcher of the future, *ASLIB Proceedings* 60 (4), 290–310.

14

The students have iPods: an opportunity to use iPods as a teaching tool in the library

Iris Ambrose

Introduction

This case study is about making the most of an opportunity, combining a traditional library introduction lesson with available mobile technology. Shepparton Campus of La Trobe University, situated in central Victoria, Australia, has a Diploma of Education programme; all students in 2010/11 were lent an iPod Touch for their year of study. The librarian needed to introduce library services to the students, as a large group, in a short period time. What better way than to make use of the iPods? The result was an easy lesson in both preparation and delivery for the librarian. Students found the session informative and entertaining. Novice iPod users were challenged to improve their skills and competent users were quickly using their iPods to navigate the library website.

iPods in libraries

One of the top ten trends identified in academic libraries in 2010 was that the 'explosive growth of mobile devices and applications will drive new services' (ACRL, 2010). University students, amongst whom ownership rates are high, expect to be able to access their institution's services easily, including the library. To meet such user expectations, librarians are being encouraged to think creatively by taking into account users' needs and preferences. Librarians making use of iPods is an example of 'leveraging the technology that their patrons use . . . to deliver robust new services without making users leave their comfort zones' (Kroski, 2008, 33).

McDonald and Thomas (2006) contend that libraries have actually done 'little to embed themselves and their resources into the everyday tools, spaces and activities important to today's learners'. They identify this as a 'disconnect' with ubiquitous hand-held access. This paper outlines one exploration of alternatives for the delivery of library information literacy skills.

The use of mobile devices for information literacy can be seen as an extension of Web 2.0, described as a 'crucial deficit area' on account of the limitations of the technology and the variety of devices owned (Godwin, 2010). Despite the perceived difficulties, other librarians are also experimenting. Cairns and Dean (2009) describe using iPods to 'deliver some of our most basic freshman level library instruction to students'. Further explorations of iPod use in the library setting include development of the library tour as a podcast targeted to a specific audience (Lee, 2006), pre-loaded iPods for loan for listening assignments, or uploading educational podcasts to iTunes University, where an increasing range of materials are available free of charge. These innovative ideas will require librarians to change their attitudes and behaviour towards mobiles in libraries, 'harnessing them rather than banning them' (Godwin, 2010, 207).

Lippincott (2010, 3) questions the goals of libraries when launching new services for mobile devices. She speculates that this may be done 'purely because a large segment of their user population owns those devices and uses them regularly'. However, in this case study, mobile devices were seen as an excellent reason to seize an opportunity because they provided additional benefits, such as 'enhancing the library's instruction program, both in the classroom and beyond', which she sees as a more acceptable goal (Lippincott, 2010, 4). The delivery method was face to face but it made use of the iPod Touches that students already had.

Background

Each of the 80 postgraduate students in the Diploma of Education (Middle Years) course received an iPod Touch at the beginning of semester one, 2010, which allowed them to download lecture podcasts, access online resources and share their own work with others. The initiative was funded by the University, with the students returning the mobile devices on completion of their diplomas. The Faculty of Education academic co-ordinator at the Shepparton Campus, Caroline Walta, is exploring more flexible ways of online learning for the students, and also ways to engage these trainee teachers with how they might be able to use mobile technology in schools.

All the students on the course need to become competent users of electronic resources from the Library. The blended learning structure has students on campus for only five weeks; online modules and teaching practicum placements account for the rest of their year. Physical access to the library is possible only for some who live locally. They must all use the virtual library for essential readings, e-book access and full-text journal databases for individual research.

Library orientations in a small campus library have become inadequate when catering for large groups. The dilemma was the need to introduce 80 students to the Library in the absence of a training room. In the past, introductory sessions have been given as a lecture, with demonstration of the Library website and searches conducted on screen, the students passively taking in and quite possibly forgetting the detail before even leaving the room. A hands-on class would be far preferable. Making use of the iPods solved this problem.

Lesson description

The Diploma of Education cohort was split into two groups of 40 in the main study area of the Library for the first introductory session at the beginning of semester. The students had received their iPods only the day before, with one workshop on how they could be used; most were novice users but were learning quickly. Enthusiasm was high and there was much sharing of knowledge. 'How do you do that sliding thing?', 'If you tap the screen it will . . .' The session consisted of the librarian explaining different parts of the Library website and suggesting sample searches in the catalogue while the students followed individually, asking questions when needed. Impressively, a few students were able to do a relevant search, place a hold on an inter-campus loan and send an e-mail to the librarian with feedback, all before the session was even completed. There were also a few students who gave up on manipulating the small screen, deciding to watch others instead of falling behind the class: 'I'll try again later in my own time.'

The second session was held during the mid-year on-campus week, mainly to demonstrate a new catalogue search feature because the students were embarking on the research required for the major philosophy essay. By this time all students were experienced iPod users. They knew what to expect and came prepared, with iPods ready to log into the Library website. Again the session aimed to do some keyword searching of the catalogue, open an e-book and find full-text journal articles. This time the searches were up to the individual, with the librarian simply giving advice and encouragement. At the conclusion of the session several students were asked if they would answer some questions about how they found both sessions.

Results

Ten students responded to the e-mailed follow-up questions; they were both novice and self-rated expert users (who also used an iPhone). Responses ranged from the succinct *sent from my iPod* message 'Great, worked, saved Piaget paper, ta' to the more descriptive 'I am unfortunately a rare user of my iPod. Not thru lack of wanting just very time poor this year. I have not had the opportunity to use my iPod for searching library catalogues as I find I do this mainly at home on the PC after the kids have gone to bed.' Library services that some of the students had used included finding out library hours, contact details, catalogue searching, requesting a hold, reading an e-book and opening full-text e-journal articles.

Asked 'What are the good things about using an iPod to access library services?' their answers included: 'the easy access', 'how quick it is to use and how it is still readable on the iPod screen', 'mobile within any Wi-Fi zone', 'I have it in the house so I can use this whilst doing housework/looking after kids'. All these responses describe the advantages of any mobile communications device.

Asked 'What are the barriers that prevent you doing these things?' their answers included: 'you have to have WiFi to connect to the internet', 'I kept losing the internet connection', 'if no WiFi access, iPod as stand alone is useless', 'I find it incredibly difficult to use as the screen is far too small to navigate through the site successfully'. These comments indicate the limitations of small hand-held devices with internet capability.

There were no positive responses to the question whether they had taught others to use an iPod. It had been considered that these students might be early adopters who would spread the word about accessing the Library through an iPod, but this was not the case with this group.

From a teaching viewpoint, the sessions, including the selection of relevant searches and a list of information important to know, were simple to prepare. The emphasis was on the electronic availability of resources because most of this group would become 'distance students' after their first two weeks on campus. There were no complaints about not being able to connect to the wireless network, which is a common problem with first-time laptop users trying to configure operating systems. Such excellent connectivity in the Library enabled the session to get off to a smooth start. As for any classroom lesson, there was a need to give very clear, explicit instructions and to be mindful of those not keeping up and those racing ahead. Mobile devices can 'foster active learning techniques for varied learning styles . . . another way of communicating our messages and services' (Godwin, 2010, 210).

Discussion

As part of the growing transition from face-to-face interaction with library users to virtual interaction, this session combined a traditional library introduction lesson with the available mobile technology to teach the same content in a way that students would be able to use after the class and throughout their course.

The mobile environment offers 'new venues for teaching digital literacy skills . . . and aids libraries in their outreach' (Vollmer, 2010). Wireless availability allows teaching to occur anywhere on campus without the need for a special classroom. The conversation with the library is improved by the ability to send an e-mail, ask a question of the reference desk or 'live chat' with a librarian through students' iPods or similar devices. Mobile library services allow students access on the move, even within the library building. A further benefit could be a reduction in queues to access library computers when students need to search the library catalogue. Better still would be the ability to access real-time information on the availability of resources through 'an interactive graphical map detailing location of specific resources in the library' (Meere et al., 2010, 65).

While the iPod is too small for some to feel comfortable using it, the technology in mobile devices is rapidly evolving. Some universities are considering introducing iPads, loaded with e-textbooks, as compulsory 'booklist' items or on loan from their faculties. If this is happening in an institution, it is an efficient and sensible use of resources for the librarians to use the devices in library instruction sessions. Asking students to bring their mobile devices (iPod, iPad or smartphone) to on-campus library introduction sessions is a way to engage students by using technology they already know.

Conclusion

As academic librarians experiment with ways to use mobile devices for information literacy, opportunities will arise where they can become involved with existing programmes. A successful strategy at the Shepparton Campus of La Trobe University made use of students' access to iPods as an efficient way to present library information and services to students. The retrieval of information on a mobile device is simple. Introducing students, whether novice or expert iPod users, to library information and services made efficient use both of space in the campus library and of the librarian's time. The students will continue to take advantage of the flexibility and connectivity of their mobile devices during the course of their study.

References

ACRL Research Planning and Review Committee (2010) 2010 Top Ten Trends in Academic Libraries: a review of the current literature, *C&RL News*, June.

Cairns, V. and Dean, T. C. (2009) Library iTour: introducing the iPod generation to the academic library. In *Proceedings of the 14th Annual Instructional Technology Conference, held at Middle Tennessee State University on 29–31 March 2009*, www.mtsu.edu/itconf/proceedings/09/MTSU_Conference_Proceedings_ Cairns_and_Dean.pdf.

Godwin, P. (2010) Information Literacy gets Mobile. In Ally, M. and Needham, G. (eds), *M-Libraries 2: a virtual library in everyone's pocket*, Facet Publishing.

Kroski, E. (2008) Library Mobile Initiatives, *Library Technology Reports*, 44 (5), 33–8.

Lee, D. (2006) Marketing 101: iPod, You-pod, We-pod: podcasting and marketing library services, *Library Administration and Management*, **20** (4), 206–8.

Lippincott, J. K. (2010) Mobile Reference: what are the questions? *The Reference Librarian*, **51** (1), 1–11.

McDonald, R. H. and Thomas, C. (2006) Disconnects between Library Culture and Millennial Generation Values, *EDUCAUSE Quarterly*, 4, 4–6.

Meere, D., Ganchev, I., Ó'Droma, M., Ó'hAodha, M. and Stojanov, S. (2010) Evolution of Modern Library Services: the progression into the mobile domain. In Ally, M. and Needham, G. (eds), *M-Libraries 2: a virtual library in everyone's pocket*, Facet Publishing.

Vollmer, T. (2010) *There's an App for That! Libraries and mobile technology: an introduction to public policy considerations*, ALA Office for Information Technology Policy, Policy Brief 3 (June),
http://ala.org/ala/aboutala/offices/oitp/publications/policybriefs/mobiledevices.pdf.

Part 3

Focus on technology

15

Mobile services of the National Library of China

Wei Dawei, Xie Qiang and Niu Xianyun

Introduction

This chapter focuses on the history of 'NLC IN Your Palm' and introduces the following public services: SMS, Wireless Application Protocol (WAP) and mobile phone offline applications. With 3G communications technology, the National Library of China (NLC) will further strengthen its research and development of mobile terminal services and resources. NLC will also strengthen the development of mobile reading using hand-held mobile reading devices. This chapter also analyses the challenges for mobile library services, including multi-platform terminals, the organization of resources, user experience and copyright protection in the mobile internet environment.

Mobile services of China's public libraries

With the development of mobile technology, promotion of mobile applications, gradual popularization of 3G and emergence of 4G, the numbers of mobile phone and smartphone users are gradually growing. As was revealed by the Ministry of Industry and Information Technology of the People's Republic of China (PRC) in the 2010 Statistical Bulletin of the National Telecom Industry, the number of mobile users had reached 859 million as of 2010, including 47.05 million 3G users. The number of mobile phone netizens reached 303 million, accounting for 66.2% of the total number of internet users. Thus, more and more users are accessing information from the internet via mobile phones and other mobile devices. The survey showed that, with the opening and popularization of 3G services offered by various operators, 46.5% of mobile phone netizens would be accessing the internet by 3G mobile phone in the following six months. In the era of the mobile internet, information access modes are developing towards personalization, targeting information and so on. The mobile internet will become the most important supplementary internet tool for information access and communication.

The development of mobile communication technology is driving libraries' mobile services. Domestic mobile library services have been increasing gradually since 2000 and entered a concentrated stage of development from 2005. Before 2007, the services were mainly based on SMS. After 2007, with the development of wireless network technology, WAP services gradually emerged, forming a complementary pattern with SMS services. In 2010, some libraries started to try out mobile phone app store services.

NLC prepared to launch mobile services in 2006. Following the pattern of mobile communications and consumer behaviours, it launched the NLC SMS platform in 2007, providing readers with basic services and information through SMS. In the same year, NLC launched the offline mobile phone service 'NLC Roaming' to meet the needs of readers. In 2008, with the opening of Phase-II of the New Library of NLC and construction of the National Digital Library, NLC has launched the NLC WAP site, enriching the modes and content of mobile services. NLC started to build mobile service resources in 2009 and commenced the construction of a mobile phone app store in 2010. So far, there are three applications in the app store. In 2010, NLC further enhanced its services to users by updating its WAP site to WAP 2.0. The new media service of NLC has developed fast, forming another service system in addition to in-library services and internet services, and having a significant influence in the library sector nationally.

Research and development achievements of NLC mobile services

'NLC IN Your Palm'

'NLC IN Your Palm' is a general name for services based on mobile phones and other mobile terminals provided by NLC, with a focus on SMS, NLC Roaming, NLC WAP, mobile reading and other services.

SMS

SMS is a basic and traditional form of mobile service, and the most widely used in the mobile communications sector. It is available on nearly all mobile phones. NLC officially launched its SMS in 2007 by incorporating SMS technology into its operations and providing a service for readers to which they can connect via their reader cards. The major functions of the service include book-return alerts, renewals, reservation arrival notifications, reporting lost reader cards, forwarding opinions and suggestions, information bulletin and others. The SMS offered by NLC is free of charge. Readers can receive the various kinds of messages issued by NLC for free, but must pay their mobile operators at normal SMS rates for sending messages to NLC's dedicated SMS number.

WAP service

The NLC WAP service (Figure 15.1), called the NLC Mobile Phone Portal, was officially launched to provide services to readers on 9 September 2008, based on WAP 1.0.

WAP service is a form of browser-based mobile service and is an important area of development for mobile services in the mobile

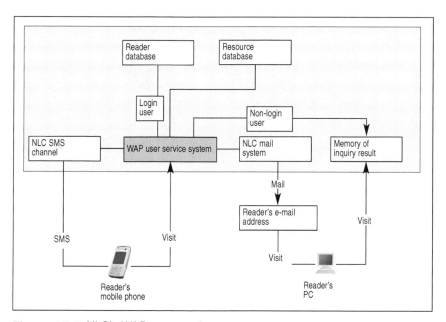

Figure 15.1 NLC's WAP user service

internet era. The NLC WAP service, called the NLC Mobile Phone Portal, was officially launched to provide services to readers on 9 September 2008, based on WAP 1.0. It was upgraded to WAP 2.0 in September 2010, improving the types of services available, mobile portal system performance and the user experience, and greatly enhanced the quality of services to readers. The mobile phone portal system uses the auto-adapter mechanism of mobile phones, and is available in simple, colour and 3G versions. It automatically selects the most appropriate version based on detection of the type of mobile phone, thus solving the interface problems that arise from differences between mobile devices.

Following a deep analysis of the services required by readers, NLC's Mobile Phone Portal provides access to resources such as Reader Service, Online Service, Reader's Guide, Wenjin Book Awards, Press Release, Resource Retrieval, etc.

Reader Service covers book renewals, book-return alerts, lending information, book reservations and arrival notifications, and other services closely related to readers.

Online Service includes online lectures, online exhibitions, online reading, book highlights and lecture notices.

NLC Mobile Phone Portal offers retrieval via the OPAC and special resources retrieval services so that readers can retrieve library resources at any time. Because it is a mobile service, retrieval and searching are free from geographical constraints and readers can access the portal from anywhere. NLC Mobile Phone Portal also provides a general description of NLC's mobile services, related software for downloading and message boards.

Wenjin Book Awards covers prize-winning books and book recommendations, books recommended by the readers of prize-winning books, votes for book awards, etc.

Offline application

NLC's offline application is a kind of client-based mobile service, NLC Roaming, launched in 2008. NLC Roaming is a navigation service that helps readers to find their way around NLC by providing reading room locations and a user-friendly guide with entertainment-like effects. It covers static navigation, dynamic navigation, service navigation, smart positioning, wireless service, reader guides and other features to enable readers to become familiar with and use NLC.

Static navigation provides floor plans of NLC's main buildings so that readers can locate their whereabouts within the buildings and find the various reading rooms.

Dynamic navigation suggests routes through the building and automatically guides readers to their desired destination.

Service navigation allows readers to search for external service outlets of NLC that they want to visit and input their current location. The system then guides them to their destination.

App store

With the emergence of new types of mobile phone such as the iPhone, the mobile phone app store has become a new mobile application model. NLC has a keen understanding of this development and is experimenting with the service model of the mobile phone app store. It has created reader service applications for the iPhone, covering OPAC retrieval, inquiries about loans, reservations, renewals, a readers' guide, information bulletin, lecture notices, etc. NLC is the first library in China to provide this type of service.

Co-operation with mobile operators

The development of mobile libraries is subject to the upgrading of mobile technology and development of the mobile internet. Mobile operators have a great technology and market advantage in terms of networks and platforms. NLC aims to promote its mobile library service using the resources provided by mobile operators. In 2009, China Mobile Limited started to build a mobile reading platform, and NLC entered into an agreement with China Mobile Limited to provide content for it. In 2010, NLC made an agreement with China Network Television (CNTV) concerning the development of a mobile TV project, which is now in progress.

Development planning for NLC mobile services

Strengthening resource development, with NLC Mobile Phone Portal as the platform

The browser-based application model is an important direction of development for mobile services in the era of the mobile internet. WAP is not going to be replaced by web, and it has a distinct user base. The China Internet Network Information Center reports that WAP accounts for more than 50% of mobile use, such as when travelling between home and office, so the mobility of WAP and its availability anywhere are important in attracting users. Most mobile phone users will continue to access WAP through their phones.

NLC's Mobile Phone Portal platform has been built for a range of resource types including pictures, audio and video, text, etc. Many resources are simply converted from original formats for access on mobile devices. In the era of mobile internet, these will not satisfy the information needs of users, so NLC must both develop its own strategies and outsource to developers in order to meet readers' needs. On the one hand, NLC needs to carry out in-house development of its own resources, such as special reports construction and rich-media resources. Also, it will pay close attention to developments by content providers and outsource projects in due course.

Mobile reading and hand-held reading devices

Mobile reading refers to browsing, watching or listening on a mobile terminal, such as a mobile phone, through wireless/mobile communications network access and information download (CEALIVE, 2010). With the emergence of mobile reading terminals such as smartphones, e-readers, the iPad and MP4, there are more and more options for readers. However, major libraries cannot introduce all the different types of mobile terminal at once to provide all the services desired by users. NLC has selected a mobile terminal that is compatible with its resources and launched a lending service for hand-held reading devices. In the future, it will introduce services for workplace users and provide services by gift, loan and many other modes, with a focus on mobile reading services.

Study via mobile library

Mobile libraries are not just about technical issues. There is also the issue of integrating library functions, management and service, i.e. extending library functions and service via the mobile internet and mobile terminals, and innovation. Further research is still required on technology capabilities, reader experience, management methods and standards for mobile libraries. NLC has already conducted research in this area and will continue and expand its research on service models, management structure, mobile terminal copyright management and other aspects of mobile libraries.

Challenges for the development of mobile services
Diversity of mobile terminals

There are many types of mobile technology, such as mobile phones, tablets and hand-held reading devices, which have different screen resolutions and operating systems. At present, the operating systems of mobile phones include iOS, Android, RIM, Symbian, Windows Mobile, Palm, Linux, OMS etc. The range of mobile phones available also covers about a dozen different screen resolutions. This diversity of mobile technologies has to be taken into consideration when providing mobile services, and somewhat increases the development costs for such services. In addition, different types of mobile devices can carry different types of content and provide different modes of service. Mobile services need to be launched that are based on the features of the different mobile technologies, such as services for hand-held reading devices, for audio and video information, for mobile reading, and other forms of service for the iPad.

Organization of resources and user experience

Organization and management of digital resources is the core of digital library development. The organization of digital resources in the internet era is well developed, but cannot be directly applied to the mobile internet. The diversity of devices for mobile internet access creates difficulties for the organization of resources because of the need to provide a sound user experience via a diverse range of technologies. In addition, solutions need to be found for a seamless service connection between the traditional digital library and mobile digital libraries. The organization of resources chiefly involves the selection of resource metadata and the indexing of object data.

Good resource organization is the basis of a good user experience, but cannot play a deciding role. Users will not use mobile technology to access information unless it is meeting their needs and they are involved in research and development of mobile products and platforms. Social and economic phenomena have given rise to the 'Experience Economy'. Based on their consumption of material goods, consumers are now more concerned about a feeling, and emotional, intellectual and even spiritual experience. The major factors influencing mobile services on mobile technologies include screen resolution of the mobile device, differences in software performance, organization of resources and the interface with users. To the extent possible, providers of mobile services need to offer the best user experience.

Copyright protection

Mobile reading is an important direction of development for mobile service and is an important means of promoting culture. As revealed by a China Internet Information Center (CNNIC) report (CNNIC, 2010), 75.4% of mobile phone netizens use mobile phone reading services, and reading via mobile phone has become the second largest application of the mobile internet. In the library field, the chief content available for mobile phone reading is open access books. These have classic content, but cater for only a small proportion of users. Newer and creative content is needed for the majority of users, and this involves copyright protection issues. The promotion of mobile reading will require libraries to work with copyright owners and mobile operators to resolve copyright protection issues.

Conclusion

Through its principle of 'user-oriented' service, NLC is paying close attention to the changing needs of libraries and library users, and is continually offering innovative services. So far, it has created a mobile service system. With the popularity of 3G and the emergence of 4G, people are becoming increasingly accustomed to using mobile devices to access information online. NLC will continue to study users' needs of and develop user-oriented access to information. In addition, it will strengthen its development of resources and carry out detailed cataloguing, so as to provide users with personalized and professional services.

References

CEALIVE (2010) *Telecommunications Industry Careers*, (12 August), www.cealive.com/2010/08/12/telecommunications-industry-careers.
CNNIC (2010) *27th Statistical Survey Report on Internet Development in China*, China Internet Information Center, July.

16

India's mobile technology infrastructure to support m-services for education and libraries

Seema Chandhok and Parveen Babbar

Introduction

The growing need for high mobility and to stay connected is the prime driver of increasing mobile use in the world today. This is true for people of different age groups and occupations. The youth segment, in particular, comprises 30% of the total mobile handset market. This segment requires high mobility and connectivity, and the same is true for business and other professionals. This innate need, coupled with the availability of handsets and connectivity at affordable prices, has triggered the growth of mobile use in India.

The mobile phone has evolved from being a mere communication device to being an access mode with an ability to tap into a plethora of information and services. This is the reason why it is now being referred to as the 'fourth screen', after cinema, television and computers.

M-resources and m-library services have become the 'pull effect', and students and researchers increasingly asking for more than just basic telephony in order to access them, driving the development of mobile value added services (MVAS). Today most of the students and faculty in universities and research institutions are seeking more from their communication devices and they are two of the prominent groups that have become increasingly reliant on mobile internet services for research, m-resources, m-educational literature searches and projects.

The growth of MVAS is based on four pillars: access devices, content, technology and infrastructure. This chapter analyses these four pillars and mobile network services in India with regard to the availability of mobile broadband; and mobile service providers with regard to MVAS and their cost-effectiveness for mobile users. The chapter will also present data about the types of mobile content and applications available through different mobile service providers and how it is used and valued by learners in different real-life study contexts. An overview of the mobile web, its potential use with Web 2.0 tools (including screencasts, podcasts, vodcasts and YouTube) and how it is beginning to be employed to support our education system, especially libraries, is also presented.

The total number of mobile subscribers in India as of July 2010 was 652.42 million. Since its liberalization in 1991, the telecoms sector has seen unprecedented growth. It is currently valued at $100 billion dollars, contributing a very significant 13% to GDP. This chapter discusses the growth of the telecoms sector, which has had a dynamic multiplier-effect on the entire economy of the nation, including education and libraries. In terms of mobile subscriptions, India is the world's second-largest wireless market after China and mobile use by about 10 million new mobile users every month. There are already crossed over 800 million subscribers in the telecoms sector (as of May 2011, there were 811.6 million mobile subscriptions in India). Fifty-nine per cent of internet users in India (23.6 million out of a total 40 million) have access to internet through their mobile devices. This provides a golden opportunity for budding Indian mobile application developers. At present around 57% of users in India's mobile sector are already using smartphones. The increasing use of mobile phones and the rapid development of India's mobile infrastructure will enhance the use of mobile technology in education and libraries. The following sections provide information on mobile use in India (see Figure 16.1 for market shares) and describe initiatives to increase the use of mobile technology in education and libraries.

Indian mobile sector

According to the new VLR (Visitor Location Register) data released by Telecom Regulatory Authority of India (TRAI) in January 2011, out of a total of 771.18 million mobile subscribers, 548.66 million (71%) were active subscribers. It is also interesting to note that Bharti Airtel, a leading global telecommunications company, has 92.63% of active subscribers, closely followed by Idea (90.34%). When it comes to

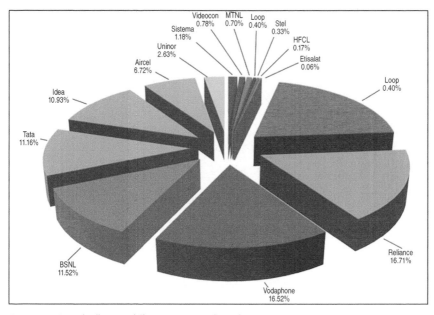

Figure 16.1 Indian mobile sector: market shares

region, Mumbai has the lowest proportion of active subscribers (59.59%), while Jammu and Kashmir has the highest (81.26%).

In January 2001 the Indian mobile sector introduced a major addition to its services: Mobile Number Portability. The new service has seen a major change in market share, as borne out by the TRAI VLR data, which shows further growth among the larger stakeholders, but not among the smaller players. Figure 16.2 on the next page shows that small providers such as Videocon and Himachal Futuristic Communications Ltd (HFCL) are actually experiencing negative growth.

Mobile value added services

MVAS are those services that are not part of the basic offer and are separately available to end-users at a premium price. Search engines and checking e-mail are the most popular uses of mobile communications in India, but the industry is constantly coming up with ground-breaking mobile technologies in order to develop new streams of revenue.

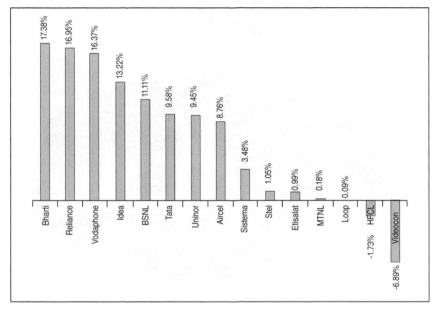

Figure 16.2 Service providers' share of net additions during January 2011

Because India has a diverse and multi-lingual population with a variety of needs the MVAS market is in a growth phase. Industry stakeholders need to ensure that the right mix of content and access is made available to the relevant user groups. With the advent of new technologies like 3G, 4G and Unstructured Supplementary Service Data (USSD), the contribution of MVAS revenues is likely to change in the near future. With the availability of vernacular content, in areas where literacy is poor, voice will gain prominence. Mobile internet access will also get a boost from the availability of cheaper data-enabled handsets, lower General Packet Radio Services (GPRS) tariffs and the provision of better Wireless Application Protocol (WAP) connectivity by telecoms operators (World of GPRS blog, 2011).

The four pillars supporting the growth of MVAS are access devices, content, technology and infrastructure:

■ **Access devices** play an important role in the use of different MVAS categories. Features like GPRS, GPS and Java, which are lacking in some handsets, are making a number of MVAS redundant. The availability of these features at affordable prices is a key factor in determining the size of the target audience. The service operators' preference for Unstructured Supplementary Service Data (USSD) as an m-payment mode, rather than GPRS, is for the simple reason that USSD is handset agnostic.

- **Content** aggregators and content owners have a larger proportion of the total MVAS cake, with a high percentage revenue share. Since they play a significant role, it is important that they generate relevant content. The content is geographically specific and not transferable across borders. Both IPR (intellectual property right) (e.g. music labels) and white label content (e.g. cricket) is available in the market. But on the other hand, much of the content being consumed is being generated for other media. Therefore a wide variety of content is available especially in the entertainment category. Investment needs to be made keeping in mind the long term benefit, and not just the short term return, on investment. One factor which impedes the content development is marketing of MVAS. Many developed content and services fail prematurely, or do not realize their full potential, because of lack of sufficient and focused marketing efforts. Currently, packaging and marketing of content is primarily in the hand of operators. These major players are not working together, claiming that the returns are not sufficient to allow this.

- **Technology**: There are two aspects to technology. One is the technology platform itself and the second is the communications technology. Technology platforms are independent of geography and transportable across borders, unlike content, which needs to have a regional flavour – for example, mobile payment platforms, IVR (Interactive Voice Response), etc. Communications technology is also independent of geography, but depends on regulatory issues, for example 3G or 4G.

- **Infrastructure**: Infrastructure requirements need to be met by providers so as to harness the potential of different technologies. Setting up infrastructure, especially in rural areas, will play a major role in the growth of MVAS.

The future of MVAS in India

The MVAS industry in India is undergoing the following structural changes and is poised to grow and contribute higher revenues to the telecoms industry:

1 Consolidation of MVAS content providers' market: Currently, the MVAS market is fragmented and consists of a large number of small content providers. Consolidation will take place in the MVAS market, leading to the emergence of a few strong content providers. This will enable content providers to command a greater share of revenues within the MVAS ecosystem.

2 Rational revenue structure: With the growth of the MVAS market in India, the revenue structure will become more rational. The revenue structure would be dependent on the value added by respective stakeholders delivering products to the end-user. This will enable the growth of the MVAS market in India.

3 Rural MVAS market: The MVAS market in India is largely dominated by the urban population, with the rural market constituting around 15% of the total market. The rural MVAS market will witness marginal growth, to around 20%, in the next two years. The growth drivers will be availability of vernacular content, entertainment services and voice-based services.

4 Differential pricing of content: The telecoms operators have already taken a step in this direction. Further, in an attempt to cater for the growing numbers of mobile subscribers in India, the telecoms operators will develop suitable pricing structures for their target markets.

5 Deployment of NGN: Introduction of a Next Generation Network (NGN) would enhance the quality of services in the MVAS market in India and users would be able to access more feature-rich services. NGN uses soft-switch technology, which is based on Packet Switching/IP phone and enables the introduction of new MVAS speed at reasonable cost. Soft switch in NGN provides basic and supplementary services called core services, while the MVAS are outside the soft switch and are application services configured through value added application servers. Generally, the application servers are interfaced with soft switch via open and flexible application programming interfaces. This will help to reduce both the launch time and the cost of new MVAS. Hence, with the introduction of NGN in India, we can expect a greater proliferation of feature-rich MVAS.

6 Number portability: The Indian government of has launched number portability in the metropolitan areas. According to the Internet and Mobile Association of India (IAMAI), this has enabled users to change service providers while retaining their mobile numbers.

Mobile social network

Mobile social networking gained popularity a few years back when smartphones introduced these applications for professionals in the business sector. But in the last three to four years, smartphones have become accessible to almost everyone. It is estimated that 30% of users in India are accessing networking sites from their mobile phones. Furthermore, the emergence of social networks such as Facebook and Twitter, and better mobile browsers that offer a rich internet experience, have also contributed to the increase in mobile data users, with social networking (24%) being one of the prime purposes for which people are using the mobile internet, after e-mail (26%) and casual browsing (28%).

According to a social media survey conducted by India Biz News & Research Services (INRS) in major cities of India, Facebook is the most widely used social software. Although more people are joining the online networking space for professional reasons, almost 90% of users join such sites in order to follow or stay in touch with their friends.

Many service providers now have unlimited data offers, making it cheaper and easier for users to use smartphones and their applications. The lower costs are important for students who want to learn and to access library materials using their mobile phones. Although growth in the use of smartphones application continues to grab the spotlight within the mobile market, the audience for mobile browsers remains large and is growing fast.

Mobile learning

The learning experience on mobile phones changes dramatically if interactivity is included. With higher bandwidths, there are opportunities for mobile learning (m-learning) to become much more interactive. For instance, video clips can be used to explain procedures, such as how to change a car tyre. Companies will make use of mobiles for training their employees.

Mobile learning includes the provision of training and learning-related content, in the spheres both of formal education and non-formal and vocational training, via mobile applications using SMS, WAP, USSD, etc. While most companies in India are largely focused on corporate learning or providing mobile alerts to notify exam timetables and results for major public examinations, there is tremendous scope for services such as language training, mobile reading, adult literacy and vocational training on specific subjects. In India, EnableM, Deltics, GCube Solutions and Tata DoCoMo are some of the players in this area. Lately, some educational institutions in India have also adopted the mobile phone as an integral part of their system because it provides a platform to support and enable the curriculum, providing students with self-tutorials and interactive tutorials in particular subjects.

Mobile technology has become significant in all spheres of life, and teaching and learning are no exception. Society is entering a knowledge-based, internet/web-driven educational economy and it is essential for people to participate in it if they want to be competitive and successful. M-learning allows on-the-go professionals to connect to training courses anytime and anywhere and can include anything from job aids and courseware downloaded to a personal digital assistant to net-based, instructor-facilitated training via mobile phone. For successful and effective m-learning, there is a need for seamless library access; support for student portfolios; a repository of shared learning objects; continuously updated class rosters; dynamic monitoring and tracking of student performance (early warning system); support for learning assessment; and enhanced student relationship management. Various m-services could include the delivery of learning materials, virtual discussions, e-lectures, e-books, m-newspapers, etc.

Several telecommunication companies (e.g. Telco) have started offering m-education services such as English lessons, dial-in tutorials, school syllabi, question sets, vocabulary and general knowledge tutorials, exam tips, exam results alerts and education for the physically disabled. These operators usually partner with value added service (VAS) companies to develop applications. At present, many leading operators are providing m-education services and applications to their subscribers. Aircel, the fifth-largest GSM player in the country, offers education-related services through its mGurujee application, which allows users to access content in the areas of engineering, management, civil services and medicine; the school syllabi of CBSE and ICSE examination boards; and skill-development, vocabulary and general knowledge tutorials. A user can subscribe to mGurujee and access learning content in practice, quiz, timed or tutorial modes.

Similarly, Tata DoCoMo provides an English Seekho service through its mobile portal, Tata Zone. It allows users to take conversational English-language lessons on their mobiles through an interactive voice response (IVR) application that guides the user through audio clips. It provides short lessons followed by interactive sessions which allow users to practise what they have learned either using the mobile keyboard or speech recognition. Reliance Communications has also been doing some work on this front, through its mobile portal RWorld. The company first launched an m-education service, called m-school, in 2003, by which teachers and parents could access databases of schools and register queries and complaints. RCom provides exam results, career counselling, etiquette and grooming sessions. It also provides English learning based on translations in rural areas, through its Grameen VAS initiative. In addition to this, Tata DoCoMo has also introduced Sparsh – an IVR-based VAS to provide sex education.

State-owned telecoms company BSNL has also started offering an English learning service for its subscribers. It has launched a spoken English programme, Learn English, which has been designed by Mumbai-based mobile content provider EnableM Technologies in association with Bangalore-based OnMobile Global. The programme teaches spoken English through simple stories and everyday situations that an individual can relate to. Subscribers can select their level of learning, based on their proficiency in the language. Daily SMS and practice tests are a part of the package, which is available in nine Indian languages. It also allows subscribers to receive a new word daily through SMS.

Apart from telecoms businesses, institutions such as the Indira Gandhi National Open University (IGNOU) have initiated basic mobile services for students across the country. IGNOU is using an SMS model, which is available in five regional sectors in India for exam alerts and has a network of 30,000 to 50,000 students. At present, most companies offering m-learning either directly or indirectly consider the industry to be very small, but they see a lot of opportunity in this area, with newer applications coming in (Bhattacharjee, 2010).

Indian libraries offering m-services

As India improves its mobile technology infrastructure, unlimited access, mobile data potential and interactive capabilities are features that are becoming prominent in m-library services. The major m-library services for library users that are provided with the help of service providers are:

1   Mobile library databases, including digitized collections of selected books, research papers, and institutional documents, abstracts of journals and periodicals, etc. Learners can turn on their phones and click on a library icon that offers them shortcuts to desired library content such as OPAC, e-books and audio books without ever having to open a web browser.

2   M-repositories: Institutional course materials, assignments, project reports and examination papers, faculty research papers, research documents, etc. are provided in mobile formats.

3   Publishers' databases: Libraries subscribe to various databases for their different client groups. These include e-books, audio books, e-journals, indexes, directories, e-resources, etc. that can be used on mobile devices. These collections can be downloaded from the library website to the user's own mobile device. Various libraries are providing authentication mechanisms for downloading to users' mobile devices.

4   Reference/enquiry services: Libraries have been providing reference service by phone for many years, and most libraries also provide reference services through a range of communication vehicles such as chat, instant messaging, texting and e-mail, which have been made easier through mobile phones. The reference desk receives calls from library users in meeting or study spaces within the library. Data from an ongoing study of Virtual Reference Services indicate that even when learners are physically in the library, they may prefer to use chat reference rather than seek out a face-to-face encounter. Again, convenience and workflow integration are important.

5   Mobile instruction: This includes the use of mobile devices for library instruction, which can be text, audio or video based.

6   Presentation and visibility: Videos and podcasts describing or promoting particular library services, covering library events and

so on are becoming more common. Often, they are made available on network-level sites such as YouTube and iTunes, where they are more visible.

7 RSS alerts: RSS is becoming pervasive. Text message and e-mail alerts are also more common. Learners can be told about events, about the status of their interactions/requests and the availability of resources. RSS feeds, widgets and Facebook applications can also be used.

8 Moblogging: Through mobile blog, learners can publish blog entries directly to the web from a mobile phone or other mobile device. A moblog helps learners to post write-ups directly from their mobile phones even when on the move, and to interact with library staff and their peer groups.

9 SMS notification: Libraries are providing SMS for a variety of purposes, such as enrolment status, examination results, launches of new programmes/services, new arrivals of resources and books, reference services, reminders of books/documents due for return, requests for lists of users' outstanding loans, document renewals, requests for summaries of outstanding fines, checking the availability of resources, etc.

10 Mobile library circulation: A hand-held circulation tool such as PocketCirc can be used that enables library staff to access the library management system on a PDA device. This wireless solution enables staff to help learners in the stacks, check out materials while off-site – such as at community or campus events – and to update inventory items while walking around the library.

11 Video conferencing: Learners can click on a mobile phone icon that initiates a video conference with a member of the library staff. With powerful services such as Skype Mobile, this has become a reality.

Conclusion

India is ready to witness mobile broadband revolution and the door is wide open for an ecosystem comprised of content providers, content aggregators, platform providers, handset makers, network equipment manufacturers, VAS providers, mobile virtual network operators and internet service providers. The Indian mobile sector is looking for new developments in education and learning models for applications, services and content that can be provided over broadband mobile networks. It is up to educators and librarians to make use of this rapidly improving and expanding mobile infrastructure.

Some challenges remain for the transformation of the physical library to m-library services. The students, learners and library users will have access to all the facilities of a physical library on their mobiles in the near future. Indian libraries are dedicated to providing m-library services to their users with the help of mobile service providers, mobile manufactures and TRAI.

References

Bhattacharjee, S. (2010) Mobile Learning Seeks Growth Curve in India, *The Mobile Indian*, 15 March, (Category: Applications),
www.themobileindian.com/.../160_Mobile-Learning-seeks-growth-curve-in-India.

World of GPRS blog (2011) India now has 771.18 Million Mobile Subscribers, (March 4),
www.worldofgprs.com/20 11/03/india-now-has-771-million-mobile-subscribers.

17

Use and user context of mobile computing: a rapid ethnographic study

Jim Hahn

Introduction

To understand the use and user context of mobile computing, a researcher from the Undergraduate Library at the University of Illinois Urbana-Champaign observed and surveyed students' use of a library iPad while riding a campus bus. Using a rapid ethnographic approach to focused qualitative data collection, the librarian questioned students about the type of journey and their destination in order to understand aspects of the idea of user context. This chapter discusses the complexities of individualized context and the implications for application development. It will give stakeholders in university campuses an understanding of users' preferences and areas for further development of mobile information access.

Work that focuses on the context of use and on users is absent from the m-libraries research literature. Specifically, design considerations should be informed by an understanding of the attributes of users' locations and the extent to which users' changing environments have an impact on information searching with mobile technologies. A recent article on urban computing points out that information access occurs in a specific place, at a specific time and on a specific device (Hansen and Grønbæk, 2010). The authors relate the importance of the mobile user being situated 'within a physical and digital context, which changes as we move about or engage in different activities' (Hansen and Grønbæk, 2010, 195). It is theorized that the physical context in which information searching and access take place will have an impact on the ways in which individuals search for information.

Usability testing, as defined by Nielsen, requires small numbers, with five to ten participants being a sufficient sample group for understanding usability concerns (1993, 169). The rapid ethnographic approach is a type of usability study. The information sought in this study was about use of the iPad in the field, and a probing approach was taken in order to gather the data quickly (Millen, 2000). Millen defines rapid ethnography as 'a collection of field methods intended to provide a reasonable understanding of users and their activities given significant time pressures and limited time in the field' (Millen, 2000, 280).

Methodology

Ten participants, all undergraduate students riding the bus, were recruited for this study, which took place during three days at the end of November and beginning of December. The rapid ethnographic approach does not necessarily produce a representative sample (Millen, 2000, 281). Rather, the probing technique allowed librarians to begin to understand what the university library might have in relation to the use of technologies like the iPad in special environments such as on the campus bus. Essentially, the value of the rapid ethnographic method was to be able to quickly form ideas for pilot library or campus services (Millen, 2000, 281).

The measure used in the study was my researcher log, shown in the Appendix to this chapter. I was interested in learning how the user's destination might have an impact on what was searched for. One aspect of the journey that I asked all students about was its routineness. Was it one that the student took every day, or was there something unique about being on the bus on that day?

I initially focused on groups of students. The literature indicates that field research on mobile computing may run into participation problems if individuals feel singled out and subsequently choose not to participate (Kellar et al., 2005, 1535). The technique of approaching two or more students worked well, but students did not always travel in groups. Since it did not drastically change the study if a student was alone and the bus was nearly empty, I did approach individual students.

Information gathering

Failing fast

In previous rapid use research my colleagues and I had worked on the premise that, in order to understand how to arrive at usable mobile access tools, we needed to uncover what does not work for mobile use (Hahn et al., 2011, 114). We call these 'fail points', and in mobile computing research we are successful if we can identify the fail points quickly. We conceptualize the rapid research process as one that will lead us to discover the fail points of mobile computing quickly, so that we can build something that does work for students.

One of the first fail points I observed was students attempting unsuccessfully to log in to the course management system. At first I considered the cause to be the keyboard interface of the iPad, which is not well suited for password entry. Neilson found the same type of usability problem when reviewing mobile apps, and wrote that mobile apps require low barriers to use and that password sign-on hinders ease of use (Nielsen, 2010). However, I later understood that the course management site does not support access by the mobile Safari browser. Even with a correct password, the iPad will not log in to the course management system because the system does not support the device's browser.

While many users considered the iPad to be useful on the bus, a few students felt that it was too large to use on the bus: it weighed too much to be taken on the bus; and another student remarked that they would not take it on the bus because they would be concerned that it might be stolen.

Mobile search successes

The majority of students (seven out of ten) reported that they thought having an iPad on the bus was useful. This student noted what she liked about using the iPad on the bus:

> The bus was bumpy but the iPad was fun and easy to use. I love the size of the screen, much better than a laptop, especially for a bus.

Another student compared the iPad interface to that of the iPhone:

It's pretty big so I probably wouldn't use it on a bus although it was easier to type on than an iPhone (internet was not functioning).

Unexpected results

Students performed searches for the dining hall menu and opening hours. Food figured in nearly half of all the students' responses. One student mentioned that they were not on a routine journey because they had 'decided to eat breakfast today' and were therefore running late and on a different bus from usual.

About half of the students interviewed were on their way to class, but not all students going to class had class-related searches. Two students going to class searched on the weather. A student who had a test wasn't quite sure what to search for; her focus wasn't so much on searching for information. She wondered aloud: 'I just don't know what to expect for the exam.' The journey to the exam did not seem to be the best time for her to search for information on the iPad. This may indicate that while students may have access to information in increasingly ubiquitous ways, they do not necessarily require ubiquitous access, since there are times when they simply don't want to search for information. It appeared that students would search for course information after class more than while on their way to classes.

Search context

I was interested in knowing how aspects of the user's journey, such as the destination and the reported routineness of the journey, might impact on the information searched for. The routine journeys were to class or dining halls. About half of the students who participated were on their way to class and the other half were on their way to the halls of residence. If a student was on her way to class, she sometimes searched for the weather, or might want to log in to the campus class management system. In one case, a journey to class meant that the student was interested in searching for information directly related to the course. The students heading to the halls of residence were interested in the dining hall hours, and then also thought that they would like to know what was on the menu for lunch.

Non-routine journeys were of interest in this study. A student who was late for class offered to participate, and reported frustration with the iPad. Another non-routine journey was by a student who had just been to the campus health clinic. Her search on the iPad was specific to her medical condition, about which she wanted to know more.

Findings

With reference to the above descriptions of the unexpected, the non-routine and early fail points, libraries will want to consider the relevance of the following points for mobile computing applications, perhaps with particular a focus on tablet computing.

Ease of authentication for resources

As noted above, the inability to log in easily to the campus course management system from the iPad was a fail point. This indicates a need for apps that can contain all campus passwords and will allow students easy, one-tap access to desired resources. If a library (or campus) app can be created with the student's necessary certifications pre-configured, then perhaps a 'keychain-like' approach could be used, such that students would not have to be prompted to log in, since the login time on an iPad might equal the amount of time available for looking at the desired information. During this study, no student spent a great amount of time using the iPad while on the bus. Most searches lasted a few minutes at the most, and the study itself never lasted longer than five minutes. Students are in transit and do not expect to be on the campus bus for long periods of time.

The integration of the library into any campus system

When designing for mobile access, the library will want to consider the extent to which any library system is able to integrate into campus-wide computing environments. It should be noted that any efficiency that systems can achieve for mobile access will make them extensible not just into other areas of campus-wide systems, but also into other, future access tools.

Forming partnerships with the information technology professionals in the residence halls

The library will want to investigate the possibility of pushing out either an RSS feed for the dining hall hours or the menu data as an XML string that can be pulled into a mobile-based application. By doing this, the computing departments of the library and residence halls will learn how to take various desired categories of information and extend them into any preferred computing environment.

Creating health data feeds for information that can be accessed in the field

While this was not necessarily a trend, one thing I noticed was that students do have questions about health concerns, and the winter months are likely be a time of increased need for health information. Also, undergraduate students in halls of residence are generally away from home for extended periods of time and taking charge of their personal health may be a new area of concern. The library cannot replace health providers, but the health resources on the web that any campus can point to could be viewed as candidates for extension to mobile computing platforms.

The varying complexities of context

Context matters, and is fraught with individualized complexity. If a student is on her way to class this does not necessarily guarantee she will want information about her class. I interviewed one student who was on her way to a test. She was more concerned with what to expect for the exam she was about to take; she could not focus on searching. Another student who was on her way to class, and late for it, disclosed that she was flustered because she was late, and because of this she could not log in to her course management site. Other students, apparently happily on their way to class, searched for the weather; students returning from class did course-related searches in chemistry and maths.

A campus mobile app

Mobile apps will be useful and used if they are designed with features that allow individualized customizability for information access. Attending to students' individual preferences and motivations for research and study means that application design in campus settings should include a high level of customizability – essentially mimicking that of smartphone interfaces – in which various modules of the campus app could be removed or added, based on the individual user's specific context or anticipated needs. Starting-points for customization could be place- and time-specific information – the dining hall menu or the opening hours of campus facilities. The value of context-specific mobile information is that it can draw attention to the various localities where users will find themselves. Campus computing departments should avoid replicating the mobile applications that already exist: weather or general map directions. These are already accessible through other applications on mobile devices.

Conclusion

This research is a start to understanding urban computing environments and possible library- or campus-developed apps for student access to information. It is experimental on two levels: first, the contextual research questions explored; and second, the method of studying information seeking, which is previously untested and is new for librarians. Future rapid ethnographic research may investigate time of day, the student's major and their year in school, so as to understand the highly individualized implications of student context and mobile information search. Additional questions for research into authentic student use of mobile technology would be to investigate students' information seeking from their personal mobile devices rather than introducing a library-provided device into the urban environment.

Appendix: research log

1 Where are you going?
2 Please use this iPad to search for any information you need before you get there.
   Note what student searches for.

3 What do you think about using the iPad on a bus?
   Prompt about: information desired, information searched for,
   information found, comments about the device, the iPad interface.
4 Is there anything special about this journey? Is it a routine trip, or
   is there something not very routine about where you are going?

References

Hahn, J., Twidale, M., Gutierrez, A. and Farivar, R. (2011) Methods for Applied
   Mobile Digital Library Research: a framework for extensible wayfinding
   systems, *The Reference Librarian* 52 (1–2), 106–16.

Hansen, F. A. and Grønbæk, K. (2010) UrbanWeb: a platform for mobile context-
   aware social computing. In *Proceedings of the 21st ACM Conference on Hypertext
   and Hypermedia, held on June 13–16*, ACM.

Kellar, M., Reilly, D., Hawkey, K., Rodgers, M., MacKay, B., Dearman, D., Ha, V.,
   MacInnes, W. J., Nunes, M., Parker, K., Whalen, T. and Inkpen, K. M. (2005)
   It's a Jungle Out There: practical considerations for evaluation in the city. In
   *Proceedings of CHI '05: extended abstracts on human factors in computing systems, held
   on 2–7 April in Portland, OR, USA*, ACM.

Millen, D. (2000) Rapid Ethnography: time deepening strategies for HCI field
   research. In *Proceedings of DIS '00, the 3rd Conference on Designing Interactive
   Systems: processes, practices, methods, and techniques*, ACM.

Nielsen, J. (1993) *Usability Engineering*, Academic Press.

Nielsen, J. (2010). iPhone Apps Need Low Starting Hurdles: Jakob Nielsen's
   Alertbox, (10 February), www.useit.com/alertbox/mobile-apps-initial-use.html.

18

Meeting the needs of library users on the mobile web

Hassan Sheikh and Keren Mills

Introduction

The Open University (OU) UK is a world-leading distance learning institution and currently has more than 240,000 students studying various undergraduate and postgraduate courses. Since 2005 a growing number of students have been accessing the University's websites on web-enabled mobile devices. In order to meet their requirements we endeavour to ensure that the content and services on the Library website are accessible and render well on smaller screens.

We developed the first mobile version of the Library website in 2007 (Sheikh and Tin, 2007), working collaboratively with Athabasca University using its ADR (Auto Detect and Reformat) software. This version was a single-column design intended to work on basic mobile phones as well on smartphones such as the Nokia N95. However, in the last couple of years our website analytics have shown an increase in numbers of visits from touch-screen phones (e.g. iPhone, HTC Android and Samsung Galaxy), which has prompted us to redesign the mobile Library website to improve usability, especially on touch-screen mobile devices. We are adapting MIT's open source Mobile Web project, which enables the website design to be optimized for three categories of small-screen devices: basic, smart and touch-screen phones.

In this chapter we will highlight some of the developments to the mobile Library website and the work being carried out during different stages of the project, specifically covering:

■ gathering user requirements for mobile library services through user feedback, focus-group consultation and website analytics

■ a technical overview of adapting and customizing MIT's open source mobile web software

■ the lessons learned and key problems when designing content/websites for smaller screens.

The importance of mobile services for distance learners

An increasing number of OU students are now equipped with access to mobile technologies such as mobile phones and small-screen hand-held devices. These mobile technologies include anything from basic phones to the latest touch-screen phones, small-screen tablets and e-book readers. The majority of these devices are web-enabled and thus provide students with always-on access to the internet. We anticipate this leading to an increasing demand for flexible content delivery and library services that can meet the perceived needs and expectations of students using such mobile devices. Mobile technology applications in education can benefit both students and educators.

Mobile Library website development

In order to meet the needs of our mobile users, we have been developing and enhancing our mobile Library services during the last few years. The current mobile version of the Library website (Figure 18.1) was first developed in 2007 in partnership with Athabasca University, Canada, using its in-house developed ADR software. Using the ADR software allowed us to make the Library website and other online content suitable for viewing on small-screen devices. The software automatically detects if a mobile device has been used to connect to the website and then renders and optimizes the content to fit appropriately on the mobile screen, by changing the layout template and style-sheet. The advantage of this approach is that the same content can be rendered on two different display models (one for normal screens, the other for smaller screens) and it has saved content authors from writing two separate versions of the same content.

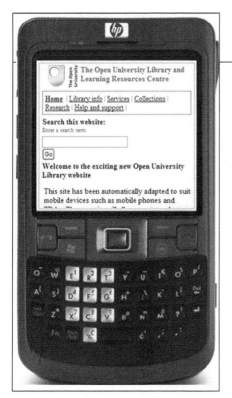

Figure 18.1 The first mobile version of the OU Library website, developed in partnership with Athabasca University

How users are using the mobile Library services and website

We have been using Google Analytics since 2007 to track OU Library website traffic (for both desktop and mobile versions) and to analyse user behaviour. The Google Analytics tool creates eye-catching graphical reports providing information on the sources of traffic (where visitors came from), what pages and areas they visited, how long they stayed on each page, how deeply they navigated into the site, where their visits ended, where they went from each page and so on.

Google Analytics statistics have revealed that the most popular areas for the Library website's mobile users are the home page, Contact Us, Opening Hours, News, Events at OU Library, e-Resources and Search the Library Collections (Figure 18.2 overleaf). These findings were also confirmed during the Arcadia research project in 2009 (Mills, 2009), when users were asked about the Library services they would most like to access from their mobile phones.

The Google Analytics data have also revealed that 66% of the total number of mobile users accessing the mobile OU Library website are using touch-screen phones such as iPhone, Android-based phones, Samsung Galaxy and Blackberry (Figure 18.3 on the next page). Based on these findings, we are currently revamping the mobile OU Library website and will focus on improving the key areas that users have been accessing most frequently.

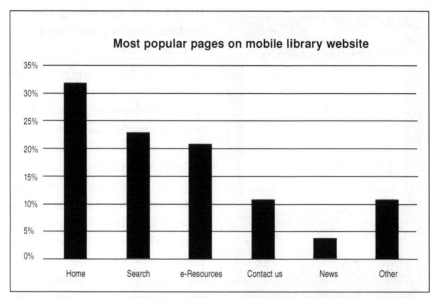

Figure 18.2 The most popular pages on the mobile OU Library website (early 2007 to September 2010)

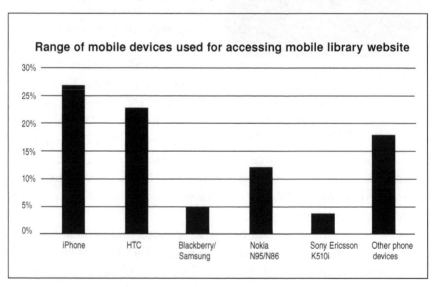

Figure 18.3 Range of mobile devices used for accessing the Library website (2010 data)

Revamping the current mobile website

One of the difficulties with a single mobile version approach using ADR software is that, although it renders well on basic web-enabled phones (and non-touch-screen smartphones such as Nokia N95), the user experience on advanced touch-screen phones such as iPhone is not optimal, on account of these phones' larger screens. For instance, the target size for links is too small to click with the finger making the user experience worse on phones with larger screens. On the other hand, the quality of videos when played on a phone with a larger screen is much better than on a basic flip phone. With the expansion of the touch-screen phone market and a large number of OU students using the larger-screen phones and devices, there is a greater demand for Library services to be developed to take advantage of the advanced capabilities and bigger screens of these devices. We are therefore revamping the mobile Library website to develop a three-devices model based on the MIT's open source mobile web project. The key areas of website redevelopment include implementation of a mobile interface for the vertical search system using Ebsco Discovery Service API; providing search recommendations; migration from our current Cold Fusion platform to a Drupal/PHP-based environment (which is officially supported by our central IT department); an improved help and support channel; and development of targeted mobile pages such as the Contact Us page, web chat plug-in and a mobile-optimized list of e-journal databases.

The current version of the mobile Library website was first developed and launched in 2007. However, based on the Google Analytics data and the feedback provided by our users, we feel that there are a number of important reasons for revamping it, including:

- One size does not fit all (too many devices and different sizes and capabilities), hence we want to develop a three-devices model based on MIT's mobile web project.
- We have found (through Google Analytics and user surveys) that users are not interested in all the pages – they want to access targeted pages only.
- It makes sense to take advantage of the advanced capabilities of touch-screen phones (larger screen, better quality of audiovisuals and location awareness).

MIT's mobile web software classifies mobile devices into the following three categories (MIT, n.d.):

■ high end – large touch screen and advanced web capabilities (iPhone, HTC, Samsung, etc.)

■ smartphones – may lack touch-screen function but have decent web capabilities and JavaScript support (Nokia N95, Windows Mobile, Blackberry, etc.)

■ low end – small screens and basic web capabilities (web-enabled flip phones).

For the redevelopment of the mobile Library website we have used a combination of Drupal and PHP for the front end development with a mySQL back-end database (Figure 18.4). The detection of device capabilities is accomplished using WURFL (Wireless Universal Resource File – a community-managed configuration file that contains information about all known wireless devices).

Figure 18.4 The new home page(s) of the OU Library website based on MIT's three-devices model mobile web project

The key mobile services in development

We are working on a number of developments, including:

■ a separate home page for each of the three versions of the mobile website

- based on the feedback and user analytics, the mobile home pages having links to the pages that users access most on mobile devices, such as opening hours, Contact Us, search, news, the calendar of library induction sessions for students, and help and support
- a list of mobile-friendly databases (e-journals) – colleagues in several libraries have tested databases on mobile phones with different capabilities and have drawn up a list (Lib-Success Wiki, n.d.) of mobile-optimized databases
- mobile implementation of 'live person' web chat so that students can get in touch with our helpdesk staff through their mobile phones
- a mobile interface to offer integrated vertical search and Amazon-style recommendations to e-resources (Figure 18.5). The recommendation work has been carried out under the RISE2 (Recommendations Improve the Search Experience) project and includes course-based, similar-article-based and similar-search term-based recommendations.

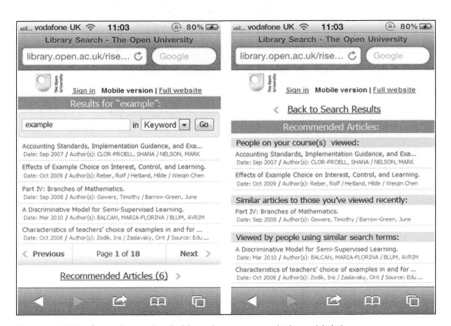

Figure 18.5 Search results (left) and recommendations (right)

Problems

The key problems we face in developing services for mobile devices are:

- too many different models of mobile device – it is difficult to keep up to date with the plethora of new mobile models coming out every week
- getting various authentication methods to work seamlessly for users to ensure simple single sign-on for access to all our subscription resources (see Figure 18.6 for an example)
- web apps offered by content providers not working with institutional login: it is difficult for the students/users to remember a large number of user credentials.

Figure 18.6 Problems authenticating via Shibboleth on mobile – too many redirects prevent students from accessing the full text of online articles

Mobile usability/design principles

Usability is a key element of the mobile development and we are trying to follow these general usability design principles:

- Know your audience and know what they are likely to want to do on a mobile device.
- Design for the type of hand-held device most likely to be used to view your web page. Web logs and user surveys can help to determine this. Our analytics show that our website is accessed far more by touch-screen phones than by other types.
- Leave space around buttons and links to ensure they have a big enough 'target size' (1 cm x 1 cm).
- Minimize the number of clicks and the volume of downloads.
- Save state – expect bad connectivity and ensure that users will be returned to the page they were on if their connection drops temporarily – especially for forms.
- Include search so that users also have the option to discover resources based on keywords.
- Reduce clicks, scrolling and time needed to complete a task.

Mobile websites/apps versus native apps

There are advantages and disadvantages to each approach. According to ComScore's Mobile Year in Review 2010 report (ComScore Inc., 2011) mobile internet browsing is far more popular than app use. However, the Nielsen Norman Group (2010) suggests that apps are better for regular or loyal users. Table 18.1 compares the three approaches.

Table 18.1 Comparison of mobile websites and apps

Mobile website	Native or proprietary app	Web app [HTML 5]
Platform independent	Platform specific	Platform independent
Use existing staff skills	Staff need new skills	Build on existing skills
Limited access to device hardware	Access to device hardware	Access to device hardware
Easy to find	Hidden in app store	Can be used on desktop
Under your control	May be regulated by app store owner	Under your control
Storage is server side	Limited memory on device	No need to develop separate versions for different platforms
Difficult to personalize	Inherently personal	
Can only be used online	Can be used offline	
	Expensive to maintain	

Conclusion and tips/recommendations

The mobile phone landscape is changing rapidly, with a sharp increase in the use of touch-screen phones. Users now expect to have a richer mobile experience because of advanced capabilities, faster internet connections and the larger screens offered by these phones. With increasing numbers of advanced mobile users, the demand for m-library services is also growing, which puts pressure on universities and libraries to improve their internet-based systems and develop multiple versions of their mobile websites, making them fit for purpose and offering a rich experience for advanced phone users, but also providing a reasonable experience for users with basic web-enabled phones.

Here are some tips and recommendations for developing mobile library websites and services:

- One size doesn't fit all – develop separate mobile versions for different categories of devices.
- Collaboration between organizations is important because it saves time. It is also important to be aware of other mobile development work taking place in your institution because it may save you some work in developing mobile-optimized templates or brand-compliant headers and footers and ensuring that your mobile library services are linked to and from other mobile services.
- Ensure that your mobile library services tie in with the institution's overall mobile service delivery strategy.
- Don't reinvent the wheel. Reuse code or use open source code!
- Use HTML5-based web apps if there isn't a specific need for a native app.
- Don't just shrink the pages, but mobilize the relevant and useful content.
- Make use of rapid and iterative development, and include regular feedback from users into the development.

Notes

1 Vertical search is a specialized web-based discovery engine that unifies search across multiple library (especially print and electronic) resources indexed in a single data harvest. The OU is

using Ebsco's vertical search system, which is also known as Ebsco Discovery Solution (EDS).

2 RISE is a project at the OU Library funded by JISC as part of the Infrastructure for Education and Research Programme, and aims to exploit the unique scale of the OU (with over 100,000 annual unique users of e-resources) by using attention data recorded by EZProxy to provide recommendations to users of the recently introduced Ebsco Discovery Solution.

References

ComScore, Inc. (2011) The 2010 Mobile Year in Review, www.comscore.com/Press_Events/Presentations_Whitepapers/2011/ 2010_Mobile_Year_in_Review.

Nielsen Norman Group (2010) Usability of Mobile Websites and Applications, www.nngroup.com/reports/mobile/.

Lib-Success Wiki (n.d.) Publishers Offering Databases for Mobile Devices, www.libsuccess.org/index.php?title=M-Libraries#Publishers_offering_ databases_for_mobile_devices.

Mills, K. (2009) M-libraries: information use on the move, Arcadia Programme.

MIT (n.d) Developing the MIT Mobile Web, www.slideshare.net/shubeta/developing-the-mit-mobile-web-presentation.

Sheikh, H. and Tin, T. (2007) Designing and Developing E-learning Content for Mobile Platforms: a collaboration between Athabasca University and the Open University. In Needham, G. and Ally, M. (eds), *M-Libraries: libraries on the move to provide virtual access*, Facet Publishing.

19

Mobile dynamic display systems for library opening hours

Keiso Katsura

Introduction

Librarians are often asked by off-site patrons about their opening days and service hours. It would be helpful for both librarians and patrons for the open/closed information of various library services to be accessible on mobile phone screens.

By using the original CGI (Common Gateway Interface) scripts and some image files, it is possible for libraries to indicate automatically whether their services are available or not. These CGI systems can be run on the mobile version of Google Maps and can also be linked with other functionalities, such as library-locating systems and RSS feeds. Using these systems, patrons can easily identify which libraries are currently open, which libraries will be open tomorrow, which libraries will be open the day after tomorrow and so on. Using other CGI scripts for library-locating systems, patrons can find libraries in specific areas and then, by means of green, red and yellow icons, like traffic signals, find out whether they are now open, closed or closing within one hour. A green icon means open, red means closed and yellow icon means closing within one hour.

Using these visual switching systems, patrons can review the information much more quickly than via systems that show the opening and service hours using static information only.

CGI systems can be applied not only to libraries but also to other cultural facilities such as museums and archives, and to business sectors, such as book stores.

Why library opening hours?

Questions most frequently asked by patrons

Queries about library opening hours are among the questions most frequently asked by off-site patrons, and they also want to access the information via their mobile phones. This was made apparent in a survey that was carried out at Cambridge University Library and the Open University Library (Mills, 2010).

Few library website pages display opening hours clearly

A few library websites have top-level pages that notify whether or not the library is open now, or is open today, whenever they are accessed. Even when this information is available on top-level pages, it is sometimes buried in the lower parts of the page or in small fonts. When library opening hours are not advertised on the top-level pages but hidden on linked pages, it takes too much time to find the information. Many libraries list their opening hours by the week. In Japan, many public libraries have separate pages detailing their annual opening calendars. When this is the case, it is easy to find out whether or not a library is open on a particular day, but it is still unclear whether or not a library is open now.

Lack of patron-centred services and co-operation

Generally speaking, the opening hours of libraries are shorter than those of other institutions (Naylor, 2005). Because libraries tend to have unanticipated closing days, patrons cannot predict when they will be closed. They sometimes have special closed periods for inventory, system replacement, renovations and so on.

Librarians often consider the library's opening hours to be of peripheral importance; they prefer to be involved in more sophisticated activities. In the area of library co-operation, librarians are much involved in establishing union catalogues, interlibrary loans and co-operative reference services – but they don't share information on these topics with others.

As for library patrons, they want to know the opening hours of multiple libraries at the same time. But librarians are satisfied with advertising only their own library's opening hours on their web pages.

Displaying opening hours on mobile screens

Library web pages that focus on library opening hours can be simple, and accessible by PCs. Smartphones with touch screens are easier to use than PCs for browsing and navigating, and libraries can easily add access for them to their web pages (Bridges, Rempel and Griggs, 2010).

Where 2.0 + When 2.0 (1.0)

Web 2.0 has morphed into Where 2.0, When 2.0 and so on. Where 2.0 is concerned with dynamic and interactive maps, such as Google Maps. Where 2.0 is characterized by the terms 'anywhere', 'ubiquity' and 'GPS'. When 2.0 is embodied in dynamic and shared calendars. Although the CGI-based displays for library opening hours are categorized as When 1.0, it is possible to embed them in Where 2.0. An even greater level of sophistication would be for library opening hours to appear in a mashed-up Web 2.0 world.

Japan as the Galapagos

Japanese mobile phone systems run on technologies that are not compatible with those of other countries (Griffey, 2010). It is thus difficult for foreign mobile phone manufacturers to enter the Japanese market. Since major mobile communication carriers are still locking their SIM cards, it is difficult for foreigners in Japan to find roaming services. The closed Japanese mobile phone environment is known, ironically, as the Galapagos.

However, this Galapagos phenomenon has yielded unique vantage points in the mobile phone environment, especially in terms of mobile teaching, in comparison with other countries. In Japan, nearly 100% of university students carry their own web-enabled mobile phones with Quick Response (QR) code-enabled cameras in the classroom. In addition, in Japan the cost of web browsing is less expensive. These factors make it possible to carry out mobile teaching in the actual classrooms.

Although this chapter is about environments that are specific to Japan, its findings can be extended to a global scale.

Dynamic display systems

Dynamic display systems are implemented by executing simple CGI scripts around icon image files. The CGI scripts can change the colours of icons, based on time and day. Figure 19.1 is an example of a CGI script. The image files consist mainly of three different colours such as green, yellow and red. The round green icon represents open, the yellow icon closing within one hour and the red icon closed. Each coloured icon is of two types: those with a transparent background change by time (in seconds) and those with a black square background change by date. The three colours are similar to traffic signals. Unfortunately, the colours cannot be displayed here. Instead, the light grey round icons indicate that the library is open or will close within one hour and the dark grey round icons indicate that the library is closed. In Figure 19.2 the round icons show the current situation.

```perl
#!/usr/local/bin/perl
#岩手県立図書館


$view1 = 'green.gif'; #開館
$view2 = 'ball-yellow.gif'; #一時間前
$view3 = 'red.gif'; #休館
$view4 = 'inventory.png'; #蔵書点検休館


($sec,$min,$hour,$mday,$mon,$year,$wday,$yday,$isdst)=localtime(time);
$mon = $mon + 1;
$mday = $mday + 2;

if ($mon == 1 && $mday <= 3) #
{
    $view = $view3;
}
elsif ($mon == 1 && $mday == 31) #
{
```

Figure 19.1 Part of the CGI script for the Iwate Prefectural Library, Japan

Figure 19.3 on page 176 is an example of pages implemented using a different CGI script. In this search system, it is possible to locate pages that focus only on library opening hours. For example, if someone carrying a hand-held device searches for opening hours in the London

area, the script retrieves some cultural facilities in the area. By clicking on the link to The British Library, the user will land on the page shown in Figure 19.2. Using this search system, information about the opening hours of cultural facilities in the UK, France, Finland and North America can be retrieved. These three parts consist of one page that can be scrolled through.

Applicability to Google My Maps

Google My Maps allows users to create personal subject maps that can be pasted into their personal pages. On Google My Maps, default markers can be replaced with tiny tailored images whose colours can be changed automatically, according to the time or date. Many libraries install Google Maps content where the marker colours cannot be changed. If colour-changeable markers are plotted onto Google My Maps, it is easy to see at a glance which libraries are open or closed, and which library services are operating.

Figure 19.2 Example of a dynamic display of library opening times (Note: these are not official British Library pages)

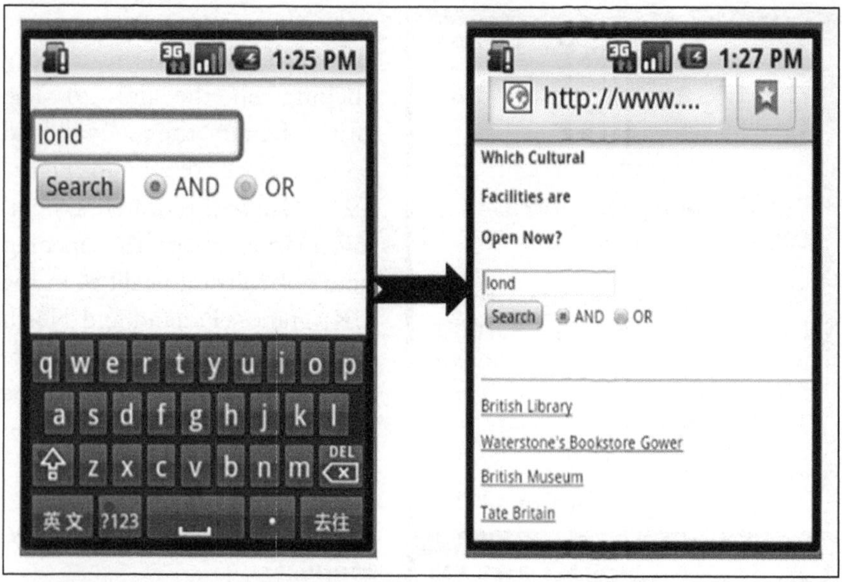

Figure 19.3 A free CGI program can search for the web pages of cultural facilities, including their opening hours. This sample shows information for the London area.

These dynamic display systems can be applied not only to opening hours but also to specific library service hours, such as chat reference services, specific reading rooms, photocopy services and soft services. Figure 19.4 was copied and pasted from Google My Maps and shows the situation for chat reference services in Europe. The map shows that a Swedish chat service is currently available. Figure 19.5 shows the open/closed status of two Japanese national libraries and about 60 major public libraries. Figure 19.6 on page 178 is an example of how the colour-changeable icons can be included in pop-up InfoWindows. By clicking on the round red icon showing that the library is now closed, an InfoWindow is opened that shows that a local library in the Osaka region is open today, tomorrow and the day after tomorrow.

Going with QR codes

The QR two-dimensional barcode reader system was invented by the Denso Wave Company, Japan, in 1994/95. It has now gained attention even outside Japan (Lippincott, 2010). Using a mobile phone camera, the QR reader app instantly recognizes a URL address. This method

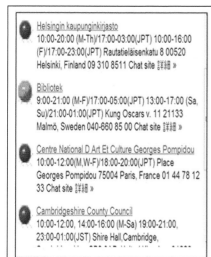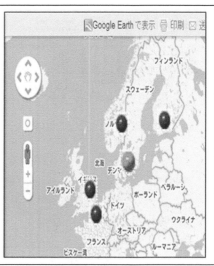

Figure 19.4 Map on Google My Maps displaying information about chat reference service hours in Europe. When the marker is green, it is possible to visit the chat site by opening the InfoWindow.

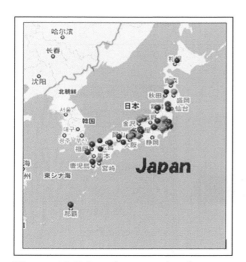

Figure 19.5 Location map showing opening/closed information for about 60 libraries in Japan

avoids the need to use the mobile phone's keyboard and is especially effective when the URL is long. The QR code system functions as a portal that takes patrons to a page showing opening hours or office hours. Figures 19.7 and 19.8 on the next page show how this process works. This example uses text images instead of colour images. By reading the URL, students can reach the web page that is linked to the office hours page. From the office hours page they can find out whether the Library Science faculty member is on or off campus, or in which class he/she is now engaged. This system can also be used in personnel management. The students can bookmark the QR code on their web-enabled cell phones for future use.

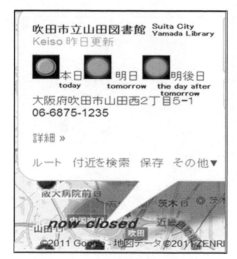

Figure 19.6 Pop-up InfoWindow showing that the Suita City Yamada Library in Osaka, Japan, is now closed, but that later today, tomorrow and the day after tomorrow it will be open

Figure 19.7 QR code attached to the entrance door of a Library Science faculty staff member's office. It leads to the web page of the staff member with his/her schedules. By using mobile phones, students can find out where the member of staff is.

Figure 19.8 By clicking the text in which the URL is embedded (left), the next page (right) shows immediately whether the faculty staff is on or off campus

Conclusion

Library users will benefit by knowing the library opening days and hours anytime and anywhere. Displaying the information with colours is easier in terms of ergonomics because the colours conveys the information instantly. Smartphones can also be used to display this information. Knowing opening days and hours ubiquitously is especially necessary for heavy users of library services. The subject is an important one, despite the silence of library science researchers on this matter.

The development of these CGI systems exposed the fact that the opening hours of libraries were shorter than those of book stores, and the frequency with which libraries are closed at irregular times.

In order to upgrade library services to the Web 2.0 era, it is important to show time-related information to patrons in real time. Many library sites, including mobile sites, lack this facility.

If display systems based on CGI scripts are replaced in the future by enhanced systems, patrons will be able to search for just those libraries that are open. Furthermore, they will be able to access library catalogues, limiting their search to those libraries that are currently open.

Web resources

Denso Wave Inc., www.denso-wave.com/en/adcd/fundamental/index.html
Google My Maps,
    http://maps.google.co.jp/maps/
Location Map of Suita City Libraries in Osaka, Japan,
    www.asahi-net.or.jp/~gb4k-ktr/suitalib.htm
Location Map of World Chat Reference Services,
    www.asahi-net.or.jp/~gb4k-ktr/chatmap.htm
Mobile Opening Hours Display for The British Library at St Pancras is created on
    Keiso Katsura's personal website,
    www.tk.airnet.ne.jp/katsura2/cgi-bin/fcf/blstpan2.htm
Search for Cultural Facilities,
    www.tk.airnet.ne.jp/katsura2/cgi-bin/fcf/wwwsrch2.cgi

References

Bridges, L., Rempel, H. G. and Griggs, K. (2010) Making the Case for a Fully Mobile Library Website: from floor maps to the catalog, *Reference Services Review*, **38** (2), 309–20.

Griffey, J. (2010) *Mobile Technology and Libraries*, Facet Publishing.

Lippincott, J. K. (2010) A Mobile Future for Academic Libraries, *Reference Services Review*, **38** (2), 205–13.

Mills, K. (2010) M-libraries: information use on the move. In Ally, M. and Needham, G. (eds), *M-Libraries 2: a virtual library in everyone's pocket*, Facet Publishing.

Naylor, B. (2005) The Battle of the Opening Hours, *Library & Information Update*, 4 (6), 14.

20

Device-independent and user-tailored delivery of mobile library service content

Damien Meere, Ivan Ganchev, Máirtín Ó Droma, Mícheál Ó hAodha and Stanimir Stojanov

Introduction

With the advent of the smartphone and the explosion in the market for mobile applications (apps), a great opportunity has emerged for the delivery of library service content to a more diverse range of users and mobile devices. While a great many individuals use iPhone or Android-based smartphones, which facilitate the delivery of these apps to users, a vast majority of the public are still left out of this particular loop. In enabling a much wider cohort of library users to access these services, the 'app store' concept has enormous potential for the enhanced delivery of learning and informational content. Whereas app stores are accessible only to certain classes of device, the aim of the system presented here is to enable any individual using a Java-enabled device to receive relevant service content tailored to their preferences and device capabilities. This chapter investigates the potential of this new architecture, encompassing mobility and the facilitation of a more personalized and contextualized information environment for library users. Also considered are the enhancements to traditional library services and practices that can be facilitated by the incorporation of these technologies, in particular the exciting possibilities they hold for a narrowing of the 'digital divide' and for revolutionizing learning in the case of distance learners and communities/populations previously starved of educational opportunities.

Throughout the 20th century and the early years of the 21st century, humanity's rate of technological advance has grown progressively swifter, with various technological innovations revolutionizing almost every aspect of life, from an individual level right up to the societal level. With the tremendous foothold that information and communications technologies (ICTs) have gained in society, it was only a matter of time before they were adopted in educational institutions and turned towards the enhancement of existing teaching and learning processes. In particular, the advent of the internet and the ability to access a seemingly infinite repository of information has altered the way people are approaching education. There have been many studies advocating greater integration of technology into education, such as Crouch and Mazur (2001), Garrison and Kanuka (2004) and Heinze and Proctor (2004). Many of these studies highlight that the use of new technologies by higher educational institutions in course delivery has become inevitable. Indeed it might be said that it is, in fact, the responsibility of educational institutions to innovate, and to promote the adoption of new technologies, facilitating enhanced learning environments for students. One of the main reasons for this is the need for educational institutions to stay relevant in this thriving knowledge age (Barone, 2005). Universities and other educational institutions play an important role in society and are tasked with preparing a highly capable and productive workforce. As such, it falls to them to be flexible enough to adapt and evolve, in order to ensure that each graduate they produce is capable of dealing with the demands of the digital era. In the past, the use of mobile devices within many educational institutions and formal learning environments has been highly restricted, with issues such as device owner and control being highlighted (Savill-Smith and Kent, 2003). Very often these devices were seen as more of a distraction for students (Roschelle, 2003), presenting an impediment to focused learning and drawing their attention away from the lecturer (Mifsud, 2002). However, as technological innovations continue unabated, the focus is now on the next step in the e-learning revolution, the shift into the mobile domain, with mobile e-learning (m-learning).

The system presented in this chapter aims to incorporate students' mobile devices into the spheres of learning, facilitating the use of a number of m-learning services to supplement traditional educational practices. Access to these services and resources is provided through intelligent wireless access points (InfoStations) located at various points throughout a university campus. This m-learning system, previously detailed in Ganchev et al. (2008), Meere et al. (2010) and Ji et al. (2011), is designed to facilitate 'anytime-anywhere-anyhow' access to learning objects, which can be adapted to suit the preferences and individual context of the user, as well as the device used to access the learning material. The first section below focuses on the infrastructure underlying the system and identifies the communication tools required to facilitate the services. The second section highlights the enhancements to traditional library services and practices that this infrastructure facilitates, in particular the possibilities for a narrowing of the 'digital divide', for greater 'social inclusion' and for revolutionizing the learning process with respect to distance learners and communities/populations previously starved of educational opportunity. The final section concludes the chapter.

System architecture

To facilitate access to mobile educational and library services (m-services) the system detailed here is based on the InfoStation paradigm, which was originally proposed to provide 'many-time, many-where' (Frenkiel and Imielinski, 1996) wireless data services. Our InfoStation-based system enables registered users to access a range of m-services through a distributed network of intelligent wireless access points or 'InfoStations' situated at key locations throughout a university campus. The system provides an ideal opportunity to enhance the m-learning experience of students, while providing a platform for other, supplementary communications services supporting the m-learning process. Some of these proposed supplementary services are dealt with in the next section. The infrastructural system is based on the concept of providing system users with wireless access (Bluetooth or Wi-Fi) to localized and contextualized services. The placement of an array of wireless portals (InfoStations) at key points throughout the campus also has a bearing on which services are accessible, because at certain locations (e.g. in the university library) the InfoStations may be specialized to provide specific location-aware services.

The InfoStation architecture involves various interoperating entities, existing throughout a three-tier infrastructure, as depicted in Figure 20.1 overleaf. These tiers incorporate the users' mobile devices, the InfoStation nodes and an InfoStation Centre:

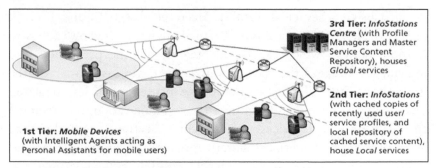

3rd Tier: *InfoStations Centre* (with Profile Managers and Master Service Content Repository), houses *Global* services

2nd Tier: *InfoStations* (with cached copies of recently used user/ service profiles, and local repository of cached service content), house *Local* services

1st Tier: *Mobile Devices* (with Intelligent Agents acting as Personal Assistants for mobile users)

Figure 20.1 InfoStation system architecture

- 1st tier – the users' mobile devices (smartphones, tablets, etc.), equipped with intelligent personal assistant agents (PAAs) that gather information about the operating environment (mobile device) and the user.
- 2nd tier – the InfoStations, satisfying the users' requests for services through Bluetooth and/or Wi-Fi wireless connections. The InfoStations maintain connections with mobile devices, create and manage user sessions, provide interface to global services offered by the InfoStation Centre and host local services.
- 3rd tier – the InfoStation Centre (the core of the overall architecture), concerned with controlling the InfoStations and overall updating and synchronization of information across the system.

Library services

Mobile technologies already play an integral role in the everyday lives of many people. Indeed, the number of mobile phone subscriptions worldwide reached 5.28 billion as of the end of 2010, according to the International Telecommunication Union (ITU, 2010). Of course, as society becomes more technologically connected and new and more innovative methods of using these technologies evolve, most aspects of individual life will develop to incorporate the use of mobile technologies. The successes of the incorporation of ICTs into education have demonstrated that when they are used effectively, new technologies can truly enhance the educational experience of students in a multitude of environments, providing a new level of freedom and a greater sense of ownership over the learning experience.

Mobile technologies have the capability to revolutionize learning and provide discontinuous rather than incremental learning opportunities in libraries and campuses worldwide. With m-learning, learning occurs across a multitude of locations, on a variety of devices, but all working towards the same goals. Lecturers, librarians and information workers are increasingly collaborating as pathfinders in this fast-growing learning environment, integrating mobile technologies and online learning. Indeed, over the past decade, the internet has spawned many innovations and services stemming from its interactive nature.

The system detailed here is used to enhance the existing library information services described below:

- Library catalogue, loans and reservations: This service builds on the existing database and cataloguing system, allowing students to access the library catalogue whilst on the move. Users can access up-to-date information as to the availability of resources and monitor when a resource is due to be returned.
- Interactive library map: This service facilitates the quick and efficient location of resources within the library. Users are provided with specific directions to the collections of materials most suitable for them (e.g. Science, Engineering, Languages) or, indeed, to the location of a very specific resource that the user may have requested through the library catalogue.
- Automated recommendations service: This service recommends library resources to users, based on their current educational context. By harnessing lists of resources accessed by previous students with similar academic interests, current students can receive recommendations of resources that may be of significant interest. This automated process will develop as more and more students from various cohorts interact with the service.

With the massive annual growth in the mobile devices market, newer and more advanced devices and technologies are continually being introduced, catering for a vast array of preferences and requirements. If various aspects of technology are not catered for, this can very easily lead to negative experiences for learners and present an impediment to them. It can be expected that within information environments such as libraries m-services will have to be accessible on devices ranging from the most resource limited up to the level of smartphones, tablets, etc. For this reason, it is vital that m-services be adapted to meet the requirements of the operating environment from which they are accessed. As proposed by Wagner (2005), the successful facilitation of m-learning requires a rich presentation layer that runs efficiently on a variety of platforms and under a variety of conditions. This adaptation and customization of m-services addresses the need for the system to serve a wide variety of devices with varying capabilities, while still delivering the service in the optimal manner.

It is not only from a basic technological standpoint that adaptation procedures are vital. With the advent of the smartphone and the subsequent explosion in the market for mobile applications (apps), the capabilities of these devices are limited only by the preferences of the user. While many individuals use iPhones or Android-based smartphones, a vast majority of the public are still left out of this particular loop. In many cases, students have access to only a relatively basic device. By enabling a much wider cohort of system users to access these services, the 'app store' concept presents an enormous potential for the enhanced delivery of both learning and informational content. However, in fulfilling this potential, the adaptation and customization of services becomes even more vital, particularly within this system. Also, whereas app stores are accessible only to certain classes of device, this system enables any individual using a Java-enabled device to receive relevant service and content tailored to their preferences.

The approach taken in the development of this system was adopted in order to narrow the 'digital divide', providing greater 'social inclusion'. The 'digital divide' refers to the gap between individuals, households and geographical areas that are at different socio-economic levels in regard both to their opportunities to access ICT and to their use of the internet for various activities (Pascual, 2003). While the term was originally coined to refer to the differing levels of access to ICT between ethnic groups, it has, over the years, come to refer to differing levels of accessibility based on socio-economic status, education, etc. The aim of this system is to facilitate access to services irrespective of the type of the mobile device or the technological knowledge of the user, and it uses the 'app store' concept to facilitate m-service delivery to an ever greater range of users and mobile devices. With an ever-diversifying student cohort entering tertiary education, the goals of this system gain even greater importance, as it is imperative that educational processes cater for all students. This system seeks to ensure that every student has equal access to (library) resources and that no student is deprived of the educational benefits of the services provided.

Conclusions

Within this chapter we have detailed a system designed to facilitate 'anytime-anywhere-anyhow' access to learning resources and adapted to suit both the preferences and the individual context of the user, as well as the mobile device used to access the learning material. As well as highlighting the underlying infrastructure, we have detailed the enhancements to facilitate traditional library services and practices through the incorporation of this infrastructure. In particular, the possibility arises for narrowing the 'digital divide' and bringing about greater 'social inclusion' and the revolutionizing of learning with respect to distance learners and communities/populations previously starved of educational opportunity.

The deployment of a system such as this has great potential for enhancing and, indeed, transforming existing library services. In this technological era, where mobility is key and when the general public have grown accustomed to using their mobile devices in so many aspects of their lives and within a multitude of information environments, the incorporation of such devices within library domains is an inevitable step. Systems such as the one presented in this chapter can provide today's librarians with another tool to deliver information effectively and efficiently to an ever-diversifying cohort of library users, regardless of the mobile devices used to access the available services. The provision of mobile services within libraries can also keep libraries relevant, particularly in an age when the internet – an almost infinite repository of information – is so freely accessible through modern 3G/4G wireless networks to a generation of young students who have grown up using these technologies.

Acknowledgments

This publication has been supported by the Irish Research Council for Science, Engineering and Technology (IRCSET) and the Bulgarian Science Fund (Research Project Ref. No. 02-149/2008).

References

Barone, C. (2005) The New Academy. In Oblinger, D. and Oblinger, J. (eds), *Educating the Net Generation*, EDUCAUSE, www.educause.edu/ir/library/pdf/pub7101.pdf.

Crouch, C. H. and Mazur, E. (2001) Peer Instruction: ten years of experience and results, *American Journal of Physics*, 69 (9), 970–77.

Frenkiel, R. and Imielinski T. (1996) *Infostations: the joy of 'many-time, many-where' communications*, WINLAB Technical Report (April).

Ganchev, I., Meere, D., Stojanov, S., Ó hAodha, M. and Ó'Droma, M. (2008) On InfoStation-based Mobile Services Support for Library Information Systems. In *Proceedings of the 8th IEEE International Conference on Advanced Learning Technologies (IEEE-ICALT '08), held on 1–5 July in Santander, Cantabria, Spain*, IEEE.

Garrison, D. R. and Kanuka, H. (2004) Blended Learning: uncovering its transformative potential in higher education, *The Internet and Higher Education*, 7 (2), 95–105.

Heinze, A. and Proctor, C. (2004) Reflections on the Use of Blended Learning. In *Proceedings of Education in a Changing Environment Conference, held on 13–14 September at the University of Salford, UK*.

ITU (2010) *The World in 2010: ICT facts and figures*, UN International Telecommunication Union.

Ji, Z., Meere, D., Ganchev, I. and Ó Droma, M. (2011) An Intelligent Framework Design for Utilization within an mLearning system. In *Proceedings of the International Conference on Data Engineering and Internet Technology (DEIT '11) held on 15–17 March in Bali, Indonesia*.

Meere, D., Ganchev, I., Ó Droma, M., Ó hAodha, M. and Stojanov, S. (2010) Evolution of Modern Library Services: the progression into the mobile domain. In Ally, M. and Needham, G. (eds), *M-Libraries 2: a virtual library in everyone's pocket*, Facet Publishing.

Mifsud, L. (2002) Alternative Learning Arenas: pedagogical challenges to mobile learning technology in education. In *Proceedings of the First IEEE International Workshop on Wireless and Mobile Technologies in Education (WMTE '02) held in Växjö, Sweden*, IEEE.

Pascuel, P. J. (2003) *e-Government*, e-ASEAN Task Force and the UNDP Asia Pacific Development Information Programme (UNDP-APDIP), www.apdip.net/publications/iespprimers/eprimer-egov.pdf.

Roschelle, J. (2003) Unlocking the Learning Value of Wireless Mobile Devices, *Journal of Computer Assisted Learning*, 19 (3), 260–72.

Savill-Smith, C. and Kent, P. (2003) *The Use of Palmtop Computers for Learning: a review of the literature*, Learning and Skills Development Agency.

Wagner, E. D. (2005) Enabling Mobile Learning, *EDUCAUSE Review*, 40 (3), 40–53.

21

Designing effective mobile web presence

Sam Moffatt

Introduction

This chapter covers the design of an effective mobile web presence so that the interaction will be friendly for users.

Every organization is different and has different sets of clients. For the University of Southern Queensland (USQ) the dominant vendor was Apple's iOS (iPad/iPhone/iPod Touch), with a 34% market share. This was followed by Nokia with 28% and Samsung and Blackberry with 10% each. HTC accounted for 7%, while LG, Motorola, Sony Ericsson and many more combined to form the final 11%. iOS as a platform is an attractive target for which to develop, given the relative homogeneity of the platform. Nokia is also large enough to merit attention, but further accounts for a wide range of device capabilities. With over 310 different mobile devices, the amount of testing and development for each one becomes a cost factor that must be considered. It is important to survey clients to find out what devices they are using and ensure that the dominant platform is supported.

Initial design

Before building a mobile web presence a few design considerations need to be taken into account. These considerations are specific to the limitations of mobile devices and the particular problems that are encountered on this platform.

Simple

Many mobile devices have a limited amount power. In some cases complex designs may take extra time to load and render, and on some simpler devices that aren't smartphones it may not be possible to display them at all. Additionally, when navigating through a website sometimes the only navigation tool the device supports is a thumbstick that allows users to navigate slowly around the page. The Nokia N series is an example of devices that rely upon a thumbstick for page navigation. Users of this platform will be alienated and frustrated by complicated designs.

A reasonable strategy for keeping the design simple is to ensure that a minimal amount of data is being sent to the client. End-users may be accessing mobile services from cellular data networks that not only are slow but also can have high associated data fees. Complicated designs with lots of graphics can slow down the transfer of the page and also increase the cost to the user of visiting the service.

Touch friendly

Touch devices are becoming the biggest players in the mobile market. The trend, popularized by the iPhone, has continued with Microsoft's release of its touch-based mobile platform (Windows Phone 7), Google adapting its Android platform to be touch capable and Nokia developing touch-capable devices on its platforms. Prior to this, many platforms featured stylus-based input, which was popular on many devices but has now faded out of use.

A touch-friendly interface is much bigger than the typical stylus or mouse-driven interface. Apple's guidelines suggest that a human finger is a 44-pixel-diameter circle. This is compared to the mouse and stylus, which allow for very precise pointing. Precision down to a single pixel means that designs that are effective for that style of pointing device may not be acceptable for a touch-based interface.

The result of this is that in order for a design to be touch-friendly extra spacing and sizing need to be considered, to aid navigation. When user-interface elements are positioned close together, the chances are increased that the user will accidentally activate the wrong option. This can be frustrating for the end-user, and could also be expensive in terms of both time and money when done over a cellular network. While many platforms have a back button in their web browser, the user may not realize what it does or may not feel confident to use it, for fear that it might set them back even further.

Another consideration for touch devices is the use of 'mouse over' gestures. On a touch device there is no concept of a 'mouse over' because there is no mouse! Designs that rely on a mouse over or use mouse overs to provide extra information will appear to be unintuitive on touch-based devices.

Easy to navigate

Mobile users are viewing information on a mobile device that has a limited screen resolution. Many devices have a portrait orientation, which means that they are taller than they are wide. In consequence of this we need to put navigation near the top of the page and have it accessible quickly. If the user has made a mistake, then the ability to go 'back' quickly and easily is helpful. Having major navigation locations, or the ability to quickly get back to the start of the site without having to click 'back' through each stage, is also an advantage. Wherever possible, navigation should be located towards the top of the page. It is sometimes tempting to locate the navigation at the bottom of the page, after the content; however, this can mean the user has to 'scroll' down to the bottom to get to where they need to be.

Navigation options also need to be simplified and clear. Long descriptions for options will often take up more space, so creativity needs to be exercised in the labelling of some options. Reducing the number of options is also important. Focus on what end-users need by creating different layouts and testing them with real end-users.

'Apparent'

The last area is perhaps the most difficult. Mobile devices need to be 'apparent', more so than their desktop counterparts, because there is no easy way to provide large amounts of help or access to a manual. This means that users need to be able to use a mobile design with only limited support and guidance. That doesn't mean that all users should be able to pick up intuitively how to use it; however, a user who is sufficiently proficient on a mobile device should be able to work out how the system functions.

Making a mobile design apparent involves aspects of the last three considerations, and also works against them. Simple designs with limited sets of options that are highlighted by their size or by other visual characteristics help to inform the user and help them to navigate.

Making an interface 'apparent' requires a lot of testing to ensure that new users can navigate through it, and feeding the results back into your development process. The more apparent the interface becomes, the greater the number of users who will be able to use it without a steep learning curve and without becoming so frustrated that they cease to be users.

On a more practical note, the foregoing implies the following:

■ the primary orientation is portrait
■ wide designs will not work well on a mobile platform
■ navigation should be located at the top.

Prototype design

An important aspect of design is to build mock-ups using either electronic systems or pieces of paper to help you examine how end-users will react to the system. This step is not particularly complicated and doesn't require the design to be exact – only to be representative of what the final design may be like. With a prototype design on paper, prospective users can be interviewed to 'road test' it. This involves the interviewer preparing pieces of paper for each aspect of the design and having the interviewee (or user) 'touch' (with their finger of course) different parts of the interface and navigate around. This can early on reveal flaws in one's understanding of how users will approach the system, and can highlight some design flaws before development begins.

Mock-ups

We will examine three designs. In the spirit of prototyping, three mock-ups are shown in Figure 21.1, one for each of the main screens.

Working from the left side of Figure 21.1, the first layout is a list navigation layout. This provides a list of items that can be selected to transition to the next screen. The second layout provides a more detail-

Figure 21.1 Design layout for mobile devices

oriented screen with some text, headings and a 'read more' link. The final layout will be the initial view for our VuFind layout, with a few navigation options at the bottom.

In all three designs common user interface elements are presented. The first is the logo and page title that sits at the head of each page. This provides consistent branding. The logo will be clickable, so as to provide a consistent location that can be used to get back to the home area. The next common features are the 'Back' and 'Help' options. Located near the top, the back button enables the user to quickly return to the last page without requiring any special knowledge of how to do so. The 'Help' option provides a link to a help screen.

Each design emphasizes simplicity in the interface and large areas to click on. Of special note is the 'read more' link, which appears in the same font size as the surrounding text. This font size is smaller than those the other major user interface elements – won't it potentially frustrate the user? The key here is that while the link itself is small, there are no other elements near it that could be selected. While other features of the layout present links that are next to each other, above the 'read more' link is text that is not linked anywhere. This means that there is sufficient space around the element for the user to click safely and not activate any other function by mistake.

Colour scheme

Before we move too far along, one of the more important aspects of a design is for the colours to be optimized. Adobe has a website called 'Kuler' (http://kuler.adobe.com), which provides a repository of colour schemes for download. If you are using Adobe graphic design tools, these colour schemes can be downloaded and integrated into applications for easy access.

Figure 21.2 shows a well-designed mobile web interface.

Figure 21.2 Example of a mobile screen with a user-friendly interface

Conclusion

The use of mobile devices in libraries requires an interface design that is user friendly and transparent to the user. This chapter has discussed techniques that should be used to build an effective mobile web presence. Similar techniques can be used to implement this sort of functionality in other systems that feature support for templates or themes. While the design presented here is not production ready or perfect, it provides a solid base or starting point for the improvement of a site for mobile users. Research should be conducted to test the ideas presented in this chapter, so that the use of mobile technology in libraries can be improved.

Conclusion

Mohamed Ally

The chapters in this book present recent research and information on the use of mobile technology in libraries. The authors of these chapters are pioneers in their respective countries and are contributing to the advancement of the use of mobile technology to transform libraries. They are laying the groundwork for the libraries of the future in the mobile revolution.

As the educational system and society change, libraries need to change too, so as to continue to provide quality service to customers. The libraries of the future will not be the same as we experience them today. They will be transformed to provide access to 21st-century services. The transformation of libraries is needed for the following reasons:

1   The new generation that is entering the education system has different expectations from previous generations. In addition to the present generation, libraries will also have to cater for future generations.
2   Technology is changing rapidly, becoming smaller, more powerful, more virtual and more user friendly. A recent news article about changes in the smartphone market had the title 'Don't Blink: you'll miss it' (Canadian National Post, 2011). This describes very well the rapid changes in technology.
3   The internet is becoming faster as we move to Internet 2 and Internet 3. This will provide fast connectivity to people, especially those in remote locations.

4 The amount of information available for access is growing fast. According to Eric Schmidt, 'every two days we create as much information as we did from the dawn of civilization up until 2003' (Schmidt, 2010). This includes user-generated information using social software.

5 The education model is changing, moving away from group-based instruction and towards individualized instruction. There are initiatives around the world to establish online and virtual educational institutions, and existing group-based institutions are moving to blended learning that includes both online and face-to-face instruction.

6 There is more emphasis on lifelong learning, since people will change careers frequently, and on the increasing availability of information. There are also initiatives around the world, such as the Millennium Development Goals, to provide education for all and to improve people's quality of life.

7 Learning for just-in-time application is taking the place of learning 'just in case' the information will be needed. Rather than sending employees to take an entire course, organizations are using e-learning and mobile learning to provide just-in-time training.

8 Learning on the go is growing in popularity. People are mobile, and they want to access information and learning materials as they travel.

9 Because of globalization, information needs to be available for 24/7 access. Learners' study patterns and the 24-hour organization provide other reasons why access to information should be available 24/7.

10 The role of teachers is changing. They will become facilitators of learning and, as a result, learners will need to access learning materials at any time and from anywhere.

11 Educational organizations and educators are making their information and learning materials available as open education resources. Again, learners will need to access learning materials at any time and from anywhere.

12 Increasing demand for informal learning, where people are learning for personal development, also requires accessibility of learning materials at any time and from anywhere.

13 Entire cities and, in some cases, entire countries have moved or are moving to wireless access. Some countries see access to the internet as a human right. Others see internet access for their citizens as an economic advantage.

14 Learning will be ubiquitous, which will require ubiquitous access to information and learning materials. Computing devices will be everywhere to give people access wherever they are located.

15 The move to Web 2.0, where users create their own content and share with others, will be followed by Web 3.0, where the machine with built-in intelligence will assemble and customize content for the learner.

The first section of the book on developing mobile services presents current information and research on services that should be provided to learners and educators using mobile technology. User-friendly and transparent services need to be provided so that customers can readily access information. Services need to cater for different cultures, languages and locations. They must also be transparent, so as to allow people with different levels of expertise and comfort with technology to access information from wherever they are located and whenever they need it. Research is needed on how to provide services for people on the move, nomadic peoples and different cultures.

The second section of the book examines the skills required by librarians in the mobile world. There is no question that the role of the librarian is changing. The librarian of the future will be more user-centric, sensitive to different cultures and comfortable using emerging technologies, and will keep up with changes in technology and function in an era of information explosion. Training programmes need to be redesigned in order to prepare librarians for the mobile world and for the future.

The third section of the book explores the use of technology to reach out to people and provide services. The mobile revolution is changing the way in which libraries provide services to people. The challenge for libraries is how to provide services to people who own different technologies. More research and development is needed to develop intelligent agents so as to provide services to people with different mobile technologies. Librarians need to work with the manufacturers of mobile technology and software developers in order to develop systems to provide a quality service to customers. Developments in the Semantic Web will build intelligence into the web in order to assemble and synthesize information-based users' needs. Librarians also need to work with educators and trainers to design information and learning materials for mobile delivery. This is critical, especially as learning materials make the shift from text to multimedia.

The information presented in this book is an excellent contribution to advancing the use of mobile technology to provide services to people. Libraries have to reinvent themselves to provide services for people in the 21st century, for the short-term mobile world and for the long-term development of emerging technologies. Current technologies will become things of the past. Social software such as Facebook and Twitter will be replaced by other, as yet unknown software, in the future. Mobile technology will be replaced by emerging technologies such as virtual devices, devices that will detect people's emotions, and devices embedded in humans and in the environment. There will be less use of textual materials and more use of multimedia.

Think back 25 years. If someone had said that users would be able to access information from libraries and obtain services using mobile devices such as mobile phones, no one would have believed it. What will libraries look like in 25 years?

Libraries need to transform themselves for the near future as well as for the distant future.

References

Canadian National Post (2011) Don't Blink: you'll miss it, (2 July),
    www.nationalpost.com/blink+miss/5038338/story.html.
Schmidt, E. (2010) Every 2 Days We Create as Much Information as We Did up to
    2003. In *Panel of the Techonomy Conference held on 4–6 August 2010 in Lake Tahoe,
    CA, USA*, http://techcrunch.com/2010/08/04/schmidt-data/.

Index

3G *see* connectivity; 3G
4G *see* connectivity; 4G

ADR *see* Auto Detect and
    Reformat
Africa 29
Amazon 16, 18, 19, 165
Android *see* touch screen;
    Android
API *see* application programming
    interface
application programming
    interface 144
apps *see* mobile apps
Athabasca University, Canada ix,
    xi, xxv, 159, 160, 161
    Press 3, 9, 169
AU *see* Athabasca University
audio xix, xxxi, 29, 36–7, 39, 52,
    54, 58–63, 65–70, 134–5,
    147–8, 163
    formats; DAISY 29; iTunes 60,
    69, 89, 96, 120, 149; MP3
    29, 39, 44, 58, 65, 67
Auto Detect and Reformat 159,
    160, 162
BeFunky 103, 108
Blackberry *see* mobile devices;
    mobile phones; Blackberry
blogging 39, 94, 134, 149
Bluetooth *see* connectivity;
    Bluetooth
broadband *see* connectivity;
    broadband

Cambridge University, UK 172
CapturaTalk 29
catalogue *see* library; catalogue
CDU *see* Charles Darwin
    University
CGI *see* computer-generated
    imagery (CGI)
Charles Darwin University,
    Australia xii, xxii, 101–6
chat 34, 79, 138, 163, 165,
    176

CHI *see* consumer health
    information (CHI)
China 34, 69, 75
cloud computing 3, 8, 85
computer-generated imagery
    (CGI) xxxiv, 170–81
computers 127
    desktop xxvi, 5, 7, 126, 127,
        167, 194
    laptop xxvi, 14, 18, 19, 55,
        137, 153
    netbook 14, 17, 18
connectivity
    3G 19, 72, 73, 75, 141, 142,
        144, 149, 189
    4G 72, 73, 142, 149, 189
    Bluetooth 34, 37, 40, 41, 185
    broadband 53, 55, 57, 69, 72,
        78, 80
    GPRS 72
    USSD 72, 76
    wireless 2, 18, 37, 55, 69, 80,
        113, 116, 126, 129, 137,
        142, 145, 147, 164, 185,
        189, 200
consumer health information
    (CHI) 34, 40
Curtin University, Australia xvi,
    14

DAISY *see* audio; formats;
    DAISY
databases 8, 25, 28, 60, 77, 79,
    87, 105, 113, 121, 124,
    135, 163, 164, 186
    vendors 60, 63, 64
Deakin University, Australia xiii,
    xxx, 13–22

desktops *see* computers; desktop
digital photography 110, 118
Diigo 110, 113
distance learning xxx, 24, 43, 158
Dolphin EasyConverter 29
Dropbox 113
Drupal 163, 164
Duke University, USA 83

e-book readers *see* mobile devices;
    e-book readers
e-books xxx, 4, 17, 28, 43–51,
    60, 77, 79, 96, 136
    formats; ePub 18, 45, 47;
    mobi 28, 29, 45; PDF 18,
    45, 47
Ebsco
    Discovery Service 163
    EbscoHost 60
Education for All *see* UNESCO
    Education for All (EFA)
    movement
EFA *see* UNESCO Education for
    All (EFA) movement
e-journals 79, 164
e-learning xi, xxvii, 8, 26, 60,
    169, 199
e-mail 15, 25, 46, 128, 136, 138
e-resources 16, 20, 27, 45, 148,
    165, 169
ethnography 152
examinations 146
Eye-Pal 29,

Facebook *see* social networking;
    Facebook
Factiva 58
Flash software 18

Flickr 94, 103
Gale Cengage 58
Google 82, 192
    Analytics 161, 163
    Maps 171, 173
    My Maps 175–7
GPRS (General Packet Radio
    Service) *see* connectivity;
    GPRS

handheld *see* mobile devices
healthcare 6, 33–8, 66
HTC *see* mobile devices; mobile
    phones; HTC
HTML 58
HTML5 168

IGNOU *see* Indira Gandhi
    National Open University
India 34, 39, 114, 139–50
Indira Gandhi National Open
    University, I • •
information technology 14, 24,
    51, 52, 78, 80, 156
information therapy 40
InfoStations xxxiv, 182–4
instant messaging 148
interactive voice response (IVR)
    143, 147
interactivity 145
internet, the 3, 6, 25, 33, 36, 38,
    40, 49, 54, 60, 66, 110,
    115, 122–3, 129, 130,
    133–5, 140, 142, 145–6,
    149, 154, 160, 168, 182,
    185–7, 197, 199
    mobile 110, 111, 129–30,
    133–6, 139, 142, 145, 167

iPad *see* mobile devices; iPads
iPhone *see* mobile devices; mobile
    phones; iPhones
iPod *see* mobile devices; iPod
Iran 34–40, 86
iTunes *see* audio; iTunes

Japan 172–180

Khan Academy 4
Kindles *see* mobile devices;
    e-book readers; Kindle
Kuler 196

La Trobe University, Australia
    119, 124
laptops *see* computers; laptop
librarians xxvi, xxvii. 2–9, 20, 40,
    51, 54–6, 75–81, 85–92,
    97, 99, 101, 103, 105, 109,
    114, 119–20, 124, 150,
    152, 157, 171–2, 185, 187,
    199–200
library
    catalogue 25–6, 45, 53, 69, 94,
    97, 105, 107, 109, 113,
    116, 121–3, 172, 179, 185
    OPAC 45–6, 48, 53–4, 132,
    139, 148
    request services 16, 23–8, 122,
    149, 184–5
    virtual 44, 121
Library 2.0 86

maps 171–7
Massachusetts Institute of
    Technology (MIT), USA
    mobile web project 159, 163–4

m-education *see* mobile learning
Microsoft Office 18
Million Dollar Book Collection 4
MIT *see* Massachusetts Institute
    of Technology (MIT), USA
m-learning *see* mobile learning
m-librarians *see* librarians
MMS *see* Multimedia Messaging
    Service (MMS)
mobile apps 39, 97, 101, 111–13,
    153, 155, 157, 167–8, 181,
    186
    Worldcat Mobile 27
mobile broadband *see*
    connectivity; broadband
mobile devices xvi, xvii, xviii, xix,
    xxv, xxvi, xxix, 4, 6, 13–22,
    28, 29, 59–67, 70, 79,
    91–100, 102, 109–15,
    116–23, 124–32, 133–9,
    141–9, 156, 158–69,
    182–90, 191–7, 201
    e-book readers 4, 16, 27, 44–9,
      98, 134; Kindle 16–17,
      18, 19, 45, 47
    iPads 17, 19, 95, 98, 101, 124
    iPods 66–71, 119–124
    mobile phones 14, 16, 27–8,
      33–40, 47, 55, 107, 129–35,
      140–9, 159–61, 165,
      172–3, 178, 200
        Blackberry 113–4, 161–2,
          164, 191; HTC 159,
          162, 164, 191; iPhone
          27–8, 33, 36, 39, 79,
          113–4, 122, 133,
          153–4, 159, 161–4,
          186, 191–2; Nokia

      159, 162–4, 191–2;
      Samsung 159, 161–2,
      164, 191; smartphones
      14, 18, 26–7, 39, 62,
      66, 69, 97, 101–2,
      104–5, 115, 134, 140,
      145, 159, 163–4, 173,
      179, 181, 184, 186,
      192
    MP3 players 62, 66, 69
    PDA 14, 33, 36, 39, 94,
      130, 149
    tablets 47, 94, 101, 135,
      160, 184, 186
mobile internet *see* internet, the;
    mobile
mobile learning 2, 8, 145, 146,
    198
mobile literacies 97–8
mobile services 76, 79–82, 86,
    97–99, 114, 129–36, 147,
    164, 168, 187, 192, 199
mobile value added services 139,
    141
mobile web 5, 6, 140, 159–60,
    163, 191, 196
moblogs *see* blogging
MP3 *see* audio; formats; MP3
MP3 players *see* mobile devices;
    MP3 players
MP4 *see* video; formats;
    MP4
Multimedia Messaging Service
    (MMS) 36–40

National Library of China
    129–37
net generation 52

netizens 129, 130, 136
NLC *see* National Library of China
Nokia *see* mobile devices; mobile
 phones; Nokia

Online Books Page 4
OPAC *see* library; catalogue;
 OPAC
open access press 3
Open Content Alliance (OCA) 1,
 4
open education resources (OER)
 5, 8
Open University, The, UK
 159–69
OU *see* Open University, The

PDA (personal digital assistant)
 see mobile devices; PDA
PHP 163, 164
Plustek BookReader 29
podcasts *see* audio; podcasts
Project Gutenberg 4

QR (Quick Response) codes 29,
 95, 98, 101–7, 113–4, 116,
 176

RefWorks
 RefMobile 27
RFID (radio-frequency
 identification) 98–9
RSS (Really Simple Syndication)
 149, 156, 171

Samsung *see* mobile devices;
 mobile phones; Samsung
screencasts *see* video; screencasts

SCU *see* Southern Cross
 University
search engines 111–2, 141
smartphones *see* mobile devices;
 mobile phones; smartphones
SMS (Short Message Service) 6,
 28, 34–40, 54–5, 115,
 129–31, 142, 147–9
social networking xxvi, 14, 39,
 145
 Facebook 39, 94, 145, 149,
 200
 Twitter 39, 89, 94, 105, 116,
 145, 200
South Africa 23–32
Southern Cross University,
 Australia 57
Stanza 97
student engagement 101–7
SurveyMonkey
surveys 29, 163, 167

tablets *see* mobile devices;
 tablets
texting 110, 148–9
touch screen 117, 18, 159–68,
 173
 Android 40, 113–4, 135, 159,
 161, 181, 186, 192
training
 end-user 16, 19, 28, 44, 66,
 102, 121, 146, 198
 library staff 8, 75, 76, 79–82
Twitter *see* social networking;
 Twitter

UNESCO Education for All
 (EFA) movement 2–10

Unisa *see* University of South
Africa
Universitat Oberta de Catalunya,
Spain 43–51
University of Bath, UK 101–2
University of Bedfordshire, UK
109–12
University of Huddersfield, UK
101–2, 109, 115
University of Illinois, USA 151
University of Southern
Queensland, Australia 191
University of Technology, Sydney,
Australia 93, 102
University of the People 3
University of the South Pacific
xxi, xxxi, 52–8
UOC *see* Universitat Oberta de
Catalunya
user generated content 1, 5
USP *see* University of the South
Pacific
USQ *see* University of Southern
Queensland
USSD (Unstructured
Supplementary Service
Data) *see* connectivity;
USSD
UTS *see* University of Technology,
Sydney

Verne, Jules, *Around the World
in 80 Days* 29
VET *see* vocational education and
training (VET) institutions

video
formats; mp4 44, 134
screencasts 97, 140
vodcasts 32, 97, 140
YouTube 94–7, 116, 140, 149
vocational education and training
(VET) institutions 75–6, 81,
102
vodcasts *see* video; vodcasts

WAN *see* wide-area network
WAP *see* Wireless Application
Protocol
Web 2.0 39, 86, 109, 120, 140,
173, 179, 199
Web 3.0 199
wide-area network (WAN) 52
Wi-Fi *see* connectivity; wireless
Wireless Application Protocol
129, 134, 142, 144
Wireless Universal Resource File
(WURFL) 164
World Digital Library 1, 4
world wide web (WWW) 37, 52
Worldcat Mobile *see* mobile apps;
Worldcat Mobile
WURFL *see* Wireless Universal
Resource File
WWW *see* world wide web

XML 156

YouTube *see* video; YouTube

Zinadoo 28